Imm

The Practical Approach Series

SERIES EDITORS

D. RICKWOOD
Department of Biology, University of Essex
Wivenhoe Park, Colchester, Essex CO4 3SQ, UK

B. D. HAMES
Department of Biochemistry and Molecular Biology, University of Leeds
Leeds LS2 9JT, UK

Affinity Chromatography

Anaerobic Microbiology

Animal Cell Culture (2nd Edition)

Animal Virus Pathogenesis

Antibodies I and II

Biochemical Toxicology

Biological Data Analysis

Biological Membranes

Biomechanics—Materials

Biomechanics—Structures and Systems

Biosensors

Carbohydrate Analysis

Cell–Cell Interactions

Cell Growth and Division

Cellular Calcium

Cellular Neurobiology

Centrifugation (2nd Edition)

Clinical Immunology

Computers in Microbiology

Crystallization of Nucleic Acids and Proteins

Cytokines

The Cytoskeleton

Diagnostic Molecular Pathology I and II

Directed Mutagenesis

DNA Cloning I, II, and III

Drosophila

Electron Microscopy in Biology

Electron Microscopy in Molecular Biology

Electrophysiology

Enzyme Assays

Essential Molecular Biology I and II

Eukaryotic Gene Transcription

Experimental Neuroanatomy

Fermentation

Flow Cytometry

Gel Electrophoresis of Nucleic Acids (2nd Edition)

Gel Electrophoresis of Proteins (2nd Edition)

Genome Analysis

Growth Factors

Haemopoiesis

Histocompatibility Testing

HPLC of Macromolecules

HPLC of Small Molecules

Human Cytogenetics I and II (2nd Edition)

Human Genetic Diseases

Immobilised Cells and Enzymes

Immunocytochemistry

In Situ Hybridization

Iodinated Density Gradient Media

Light Microscopy in Biology

Lipid Analysis

Lipid Modification of Proteins

Lipoprotein Analysis

Liposomes

Lymphocytes

Mammalian Cell Biotechnology

Mammalian Development

Medical Bacteriology

Medical Mycology

Microcomputers in Biochemistry

Microcomputers in Biology

Microcomputers in Physiology

Mitochondria

Molecular Genetic Analysis of Populations

Molecular Neurobiology

Molecular Plant Pathology I and II

Monitoring Neuronal Activity

Mutagenicity Testing

Neural Transplantation

Neurochemistry

Neuronal Cell Lines

Nucleic Acid and Protein Sequence Analysis

Nucleic Acid Hybridisation

Nucleic Acids Sequencing

Oligonucleotides and
  Analogues

Oligonucleotide Synthesis

PCR

Peptide Hormone Action

Peptide Hormone Secretion

Photosynthesis: Energy
  Transduction

Plant Cell Culture

Plant Molecular Biology

Plasmids

Pollination Ecology

Postimplantation
  Mammalian Embryos

Preparative Centrifugation

Prostaglandins and Related
  Substances

Protein Architecture

Protein Engineering

Protein Function

Protein Purification
  Applications

Protein Purification Methods

Protein Sequencing

Protein Structure

Protein Targeting

Proteolytic Enzymes

Radioisotopes in Biology

Receptor Biochemistry

Receptor–Effector Coupling

Receptor–Ligand
  Interactions

Ribosomes and Protein
  Synthesis

RNA Processing

Signal Transduction

Solid Phase Peptide
  Synthesis

Spectrophotometry and
  Spectrofluorimetry

Steroid Hormones

Teratocarcinomas and
  Embryonic Stem Cells

Transcription Factors

Transcription and
  Translation

Tumour Immunobiology

Virology

Yeast

Immunocyto-chemistry

A Practical Approach

Edited by
JULIAN E. BEESLEY
Department of Pharmacology, Wellcome Research Laboratories,
Beckenham, Kent

IRL PRESS
—at—
OXFORD UNIVERSITY PRESS
Oxford New York Tokyo

Oxford University Press, Walton Street, Oxford OX2 6DP

Oxford New York Toronto
Delhi Bombay Calcutta Madras Karachi
Kuala Lumpur Singapore Hong Kong Tokyo
Nairobi Dar es Salaam Cape Town
Melbourne Auckland Madrid
and associated companies in
Berlin Ibadan

Oxford is a trade mark of Oxford University Press

A Practical Approach 🔸 *is a registered trade mark*
of the Chancellor, Masters, and Scholars of the University of Oxford
trading as Oxford University Press

Published in the United States
by Oxford University Press Inc., New York

© *Oxford University Press, 1993*

A catalogue record for this book is available from the British Library

Library of Congress Cataloging in Publication Data
Immunocytochemistry: a practical approach/edited by Julian E.
Beesley.
p. cm.—(The practical approach series)
Includes bibliographical references and index.
1. Immunocytochemistry—Methodology. I. Beesley, Julian E.
II. Series.
[DNLM: 1. Histocytochemistry. 2. Histological Techniques.
3. Immunochemistry. QW 504.5 I331]
QR187.I45I48   1993   616.07'583—dc20   92–48716
ISBN 0–19–963270–7 (h/b)
ISBN 0–19–963269–3 (p/b)

Typeset by Cambrian Typesetters, Frimley, Surrey
Printed in Great Britain by
Information Press Ltd, Eynsham, Oxford

Preface

THE specificity and reliability of the antigen–antibody reaction have enabled immunocytochemistry to flourish for nearly 50 years and to contribute significantly to the life sciences. Today immunocytochemical techniques are widely employed in all areas from research to diagnostic laboratories. A significant advantage of immunocytochemistry is that it enables the microscopist to delve further than a morphological analysis of the sample in order to identify, and in some cases quantify, the antigen present in a specimen. Many antibodies and probes are commercially available and the technique can be performed in any routine microscopy laboratory without the need for additional sophisticated equipment. Although immunocytochemistry has long been recognized as an histological technique, increasing interest during the last 10 years has stimulated its application in electron microscopy. This in turn has precipitated new methods, such as the silver enhancement technique for light microscope application.

The art of immunocytochemistry is to unite immunology with microscopy. A specific antibody is needed and the specimen must be prepared in such a manner as to preserve the reactivity of the antigen. Antibodies have been described in two companion volumes in this series. This book will assume the possession of a suitable antibody and will describe how to prepare a specimen and how to use the antibody to achieve successful immunolabelling. I have attempted to approach the task as a logical sequence of questions which will lead the reader from general principles, through standard techniques, to more specialized applications of the antigen–antibody reaction. In doing so I have necessarily been selective as to the emphasis I have given to the available immunocytochemical techniques. I hope to inform the reader of the techniques which are routinely in use and to give an insight into the further potential of these techniques should the need arise. I further hope that this addition to the series will be of practical help to those scientists, students, and technicians needing either guidance on a particular problem or an overview of the methods for immunolabelling.

Beckenham
January 1993

J. E. B.

Contents

List of contributors xv

Abbreviations xvii

1. Introduction 1

Julian E. Beesley

  1. Introduction 1
  2. Immunocytochemistry 2
  3. History of immunocytochemistry 3
    Histology 3
    Electron microscopy 3
  4. Aims of the book 4

  References 5

2. Immunocytochemical avenues 7

Julian E. Beesley

  1. Introduction 7
  2. The antibody 7
    Polyclonal sera 8
    Monoclonal sera 9
  3. Specimen preparation 9
  4. The microscopically dense marker 11
  5. Immunolabelling techniques 11
  6. Validation of the result 12
  7. Conclusions 13

  References 13

3. Immunolabelling techniques for light microscopy 15

Peter Jackson and David Blythe

  1. Introduction 15
    Definition 15

2. Immunolabelling methods 15
  Direct method 16
  Two-step indirect method 16
  Protein A method 17
  Unlabelled antibody methods 17
  Avidin–biotin methods 18
  Immunogold methods 20
3. Chromogens 21
4. Fixation 22
5. Histological section preparation technique 23
  Paraffin sections 23
  Frozen sections 24
  Resin sections 26
6. Cytological preparation techniques 29
  Preparative methods 30
7. Immunolabelling techniques 31
8. Alternative substrates for immunoperoxidase techniques 38

  Appendix 40

  References 41

4. Immunolabelling techniques for electron microscopy 43

A. The post-embedding techniques 43

Paul Monaghan and David Robertson

1. Introduction 43
2. Light microscope characterization of reagents 44
3. Resin embedding techniques 45
  Ambient temperature 45
  Progressive lowering of temperature 46
  Freeze-substitution 51
  Immunocytochemistry on resin-embedded material 55
4. Cryosection techniques 60
  The preparation of thawed ultrathin cryosections 60
  Immunocytochemistry on cryosections 66
5. Conclusions 68

B. The pre-embedding techniques 69

Julian E. Beesley

6. Introduction 69

7. Immunolabelling techniques   71

C. The immunonegative stain technique   72
Julian E. Beesley

8. Introduction   72
9. Immunolabelling technique   74

References   75

5. Special preparation methods for immunocytochemistry of plant cells   77

Mohan B. Singh, Philip E. Taylor, and R. Bruce Knox

1. Introduction   77
2. Nature of antigens   77
     Polyclonal and monoclonal antibodies   77
     Protein antigens   78
     Recombinant proteins as antigens   78
     Carbohydrates and glycoproteins as antigens   79
3. Standard protocol for plant cells   80
     Pre-embedding method   80
     Post-embedding method   81
4. Special protocols for plant cells   86
     Highly vacuolate or fragile cells   86
     Desiccated cells   88
     Organelles   96
     Cell walls   97
     Cuticles, surface waxes, and exudates   97
5. Localization of multiple antigens in the same tissue section   98
6. Localization of antigens in transgenic plants   100

Acknowledgements   100

References   100

6. Multiple immunolabelling techniques   103

Julian E. Beesley

1. Introduction   103

2. Selection of the technique   104
   Light or electron microscopy   104
   Species of primary antibody   104
   Position of antigenic sites   104
   Immunolabelling techniques   104
   Controls   105

3. Light microscopy   105
   Choice of probes   105
   Immunolabelling techniques   108

4. Electron microscopy   116
   Choice of probes   116
   Immunolabelling techniques   118

5. Conclusions   123

   References   124

7. Techniques for image analysis   127

Nicholas G. Read and Pauline C. Rhodes

1. Introduction   127

2. Instrumentation   127
   Image analysers   127
   Cameras   131
   Microscopes   132

3. Basic steps   132
   Planning   132
   Optimizing the starting image   134
   Optimizing the analysis system   137

4. Automatic image analysis   138
   Grey level image enhancement   139
   Segmentation and production of the binary image   139
   Object and field selection   141
   Binary image editing/processing   141
   Measurement specification   142

5. Image analysis techniques   143

6. Measurement of fluorescence intensity or brightness   147

7. Image analysis of colloidal gold for electron microscopy   148

8. Concluding remarks   149

   References   149

8. Correlative video-enhanced light microscopy, high voltage transmission electron microscopy, and field emission scanning electron microscopy for the localization of colloidal gold labels   151

Ralph M. Albrecht, Scott R. Simmons, and James B. Pawley

1. Introduction	151
2. Preparation of colloidal gold probes	152
Preparation of colloidal gold spheres	152
Preparation of protein–gold conjugates	155
3. Imaging colloidal gold by light microscopy	159
The colloidal gold markers	159
Design of the light microscope for video-enhanced differential interference contrast observations	161
4. Imaging colloidal gold by electron microscopy	162
5. Labelling with colloidal gold	164
6. Summary	173
References	174

9. *In situ* hybridization   177

J. T. Davies

1. Introduction	177
What is *in situ* hybridization?	177
Why do it?	178
Practical considerations	178
2. Sample preparation	182
Fixation and embedding	182
Attachment of tissue sections	182
Checking retention of target nucleic acids	183
Pre-treatments	184
3. Types of probe—a comparison	185
4. Labelling of nucleic acid probes	186
Isotopic and non-isotopic—a comparison	186
Controlling probe length	188
5. Hybridization *in situ*	190
Introduction	190

Contents

Solutions   190

Hybridization to mRNA target using digoxigenin-labelled dsDNA probe   191

Isotopic (^{35}S) mRNA localization   192

Non-isotopic (fluorochrome) chromosomal localization   194

In situ hybridization for electron microscopy   196

6. Visualization of bound probe   198

Localization of bound non-isotopically labelled probe   198

Microautoradiography   200

7. Controls   200

8. Trouble-shooting   201

9. Conclusions—the future   203

Acknowledgements   203

Appendix: Brief explanation of terms commonly used for in situ hybridization   203

References   204

10. Problems and solutions   207

Susan Van Noorden

1. Introduction   207

2. Background problems   208

Possible causes and remedies   208

3. Method problems   218

Common errors   218

Method factors affecting immunolabelling   221

4. Specificity   231

Establishment of specificity   231

5. Multiple immunolabelling   237

6. Conclusions   237

References   238

Appendix. Suppliers of specialist items   241

Index   245

Contributors

RALPH M. ALBRECHT
Department of Animal Health and Biomedical Sciences, University of Wisconsin, 1655 Linden Drive, Madison, WI 53706, USA.

JULIAN E. BEESLEY
Department of Pharmacology, Wellcome Research Laboratories, Beckenham, Kent BR3 3BS, UK.

DAVID BLYTHE
YRCO Laboratory, University of Leeds, Department of Pathology, Leeds LS2 9JT, UK.

J. T. DAVIES
Department of Biological Sciences, University of Durham, Science Laboratories, South Road, Durham DH1 3LE, UK.

PETER JACKSON
Research Histology, University of Leeds, Department of Pathology, Leeds LS2 9JT, UK.

R. BRUCE KNOX
School of Botany, University of Melbourne, Parkville, Victoria 3052, Australia.

PAUL MONAGHAN
Institute of Cancer Research, Haddow Laboratories, 15 Cotswold Road, Sutton, Surrey SM2 5NG, UK.

SUSAN VAN NOORDEN
Histopathology Department, Royal Postgraduate Medical School, Du Cane Road, London W12 0NN, UK.

JAMES B. PAWLEY
Integrated Microscopy Resource, University of Wisconsin, 1675 Observatory Drive, Madison, WI 53706, UK.

NICHOLAS G. READ
Department of Pathology, Wellcome Research Laboratories, Beckenham, Kent BR3 3BS, UK.

PAULINE C. RHODES
Department of Pathology, Wellcome Research Laboratories, Beckenham, Kent BR3 3BS, UK.

Contributors

DAVID ROBERTSON
Institute of Cancer Research, Haddow Laboratories, 15 Cotswold Road, Sutton, Surrey SM2 5NG, UK.

SCOTT R. SIMMONS
Department of Animal Health and Biomedical Sciences, University of Wisconsin, 1655 Linden Drive, Madison, WI 53706, USA.

MOHAN B. SING
School of Botany, University of Melbourne, Parkville, Victoria 3052, Australia.

PHILIP E. TAYLOR
School of Botany, University of Melbourne, Parkville, Victoria 3052, Australia.

Abbreviations

ABC	avidin–biotin complex
ADC	analogue/digital converter
AEC	3-amino-9-ethylcarbazole
AMCA	7-amino-4-methylcoumarin
AOD	average optical density
APES	aminopropyltriethoxysilane
APAAP	alkaline phosphatase–anti-alkaline phosphatase
BPP	benzoyl peroxide paste
BSA	bovine serum albumin
BSE	backscattered electrons
CID	charge-injection device
CCD	charge-coupled device
CN	4-chloronaphthol
DAB	3,3-diaminobenzidine tetrahydrochloride
DABCO	1,4-diazobicyclo-(2,2-2)-octane
DAMC	diethylaminocoumarin
DIC	differential interference contrast light microscope
DMF	dimethylformamide
DMP	2,2-dimethoxypropane
DMSO	dimethylsulphoxide
ELISA	enzyme-linked immunosorbent assay
EM	electron microscopy
FCS	5% foetal calf serum
FITC	fluorescein isothiocyanate
HSI	hue, saturation, intensity
HVEM	high voltage electron microscope
Ig	immunoglobulin
LM	light microscopy
LN2	liquid nitrogen
LVSEM	low voltage high resolution scanning electron microscope
NGS	normal goat serum
PAP	peroxidase–anti-peroxidase
PBS	phosphate-buffered saline
PBSA	PBS+1% bovine serum albumin
PCNA	proliferating cell nuclear antigen
pf	paraformaldehyde
PLP	periodate–lysine–paraformaldehyde
PLP/D	periodate–lysine–paraformaldehyde dichromate
PVA	polyvinyl alcohol
PLT	progressive lowering of temperature
PVP	polyvinyl pyrolidone
RGB	red, green, blue
RIA	radio-immunoassay

Abbeviations

StrABC/HRP	streptavidin–biotin/horseradish peroxidase complex
TBS	Tris-buffered saline
TBST	TBS + 1% Triton X
TEM	transmission electron microscope
TRITC	tetramethylrhodamine–isothiocyanate
UV	ultraviolet
VDIC	video-enhanced differential interference contrast light microscope

1

Introduction

JULIAN E. BEESLEY

1. Introduction

Between 1680 and 1723, Leeuwenhoek used a simple microscope, capable of 30 to 200 times magnification and a resolution of approximately 1 µm, to observe insects and large bacteria. Compound microscopes evolved by the addition of a second lens and these have been fundamental in the development of the biological sciences. Abbé, in 1876, hypothesized that the theoretical maximum resolution of the light microscope (0.2 µm) was limited by the wavelength of light, and since 1900 the major advances in light microscopy have been mainly in exploiting the various capabilities of light microscope techniques and in refining methods of illumination and promoting contrast (1).

Alternatively, electrons can be used to form images, and the first electron microscopes were described in 1932. The wavelength of an electron is much shorter than that of a light ray and therefore much better resolution can be obtained with an electron microscope (0.2 nm) than with a light microscope. Because of problems with specimen preparation, the useful spatial resolution attainable for biological specimens is generally no better than 1.5–2.0 nm (2).

Light and electron microscopes have diversified and increased in complexity, often being associated with sophisticated computerization, and the instruments of today are far removed from the humble beginnings initiated by Leeuwenhoek. Early microscope research was limited to descriptions of biological structure, leaving the observer to speculate on the physiological status of the cells. In the last 20 years, however, there has been a desire to associate structural detail with chemical information and light and electron microscope techniques have been developed to investigate both the ionic and protein content of cells. If we deny ourselves access to the tremendous potential of these techniques we are denying ourselves the opportunity of transcending speculation of the physiological state of the cells judged by morphological criteria, to describing not only the presence and movement of specific proteins and ions within individual cells but also intimately linking this information with cell structure. Microscopical analysis can therefore be used to obtain a unique view of changes in individual cells, unlike biochemical analysis which assembles information concerning gross changes in cellular content of a population of cells. Immunocytochemistry is the most widely used and readily applicable of these new techniques.

2. Immunocytochemistry

Immunocytochemistry is the union of microscopy and immunology. Immunocytochemistry exploits the very specific binding exhibited by an antibody for its antigen. The reaction site can be identified in a specimen and associated with specimen structure by attaching a microscopically dense probe to the antigen–antibody complex. Proteins, carbohydrates, nucleic acids, lipids, and many other naturally occurring and synthetic compounds can act as successful antigens for raising antibodies, and consequently there is a wide variety of molecules which can be correlated with specimen structure. Microscopical analysis by immunocytochemistry can therefore be used to obtain a unique view of macromolecular changes within individual cells. If the antibody is specific, immunocytochemistry can be performed on virtually any biological specimen provided (i) the antigen remains intact and able to bind the antibody and (ii) the specimen preparation technique exposes the antigen for immunolabelling by the antibody. The resolution of the technique is limited by the length of the immunological reagents linking the antigens to the dense marker. This length is commonly less than 15 nm, which is irrelevant for light microscopy and sufficiently small to be useful for electron microscope studies. Immunocytochemistry is extremely widespread in use since it is very simple to perform once an antibody has been identified and requires no specialized laboratory equipment other than that found in a routine microscope laboratory.

Immunocytochemistry is commonly used for light and electron microscope studies. Antigens in any position within the cell can be localized. For instance, cells secreting an antigen can be distinguished from non-secreting cells; the distribution of external plasma membrane antigens can be related to either the external architecture of the cell or internal structure; internal cytoplasmic antigens can be localized with reference to cell membranes or organelles. Techniques have been developed for multiple immunolabelling, image analysis, and quantification of immunolabelling. A later development has been the development of *in situ* hybridization techniques with which gene transcripts can be correlated either to individual cells or intracellular localizations. The potential of immunocytochemistry is therefore high and it has been used in almost every area of biological research from routine diagnostic work through to complex research projects.

2

3. History of immunocytochemistry

3.1 Histology

Coons *et al.* (3) bound a fluorescent marker to an antibody and used the complex to identify antigens in tissue sections. Despite their relative age, the fluorescent probes are still widespread in use and with the advent of confocal microscopy, are experiencing a renaissance. The basic immunocytochemical philosophy described by Coons *et al.* (3) of an antibody linked to a microscopically dense marker, has not altered, but the microscopically dense marking system has been developed for application to a wide range of histological and electron microscope techniques.

Nakane and Pierce (4) and Avrameas and Uriel (5) covalently linked the enzyme peroxidase, with the second antibody for use in histology. The peroxidase can be visualized by development with one of several different substrates (6) to produce a brown, blue, or yellow reaction product. An increased sensitivity of the immunoperoxidase technique was achieved with the unlabelled antibody peroxidase–anti-peroxidase (PAP) technique. The marker is a peroxidase complex of three peroxidase molecules associated with two anti-peroxidase molecules (7). This technique gives a high signal, since three peroxidase molecules are associated with each antigen. Alkaline phosphatase (8) can be used in a similar manner. The peroxidase and alkaline phosphatase techniques are now firmly established and are widely applied. Other enzyme markers such as β-galactosidase (9) and glucose oxidase (10) have been described but are not widely used.

Alternatively, colloidal gold probes can be used for histological immuno-cytochemistry. Individual gold probes are not easily visualized with the light microscope, immunolabelling being detected as a pale red coloration in very lightly stained preparations (11). Optimum detection of colloidal gold on histological specimens is achieved by enhancing the size of the gold spheres by deposition of metallic silver (12) and observation with bright field, epi-illumination, or dark field optics.

3.2 Electron microscopy

The first immunocytochemical technique described for electron microscopy was the immunoferritin technique (13). Post-embedding immunolabelling with ferritin is not successful; its main use therefore has been for pre-embedding immunocytochemistry (7). Although still occasionally employed, its widespread use has been hampered by the degree of non-specific background labelling encountered, its small size, and insufficient density.

The immunoperoxidase and PAP techniques have been adapted for electron microscopy. For this, the enzyme is developed with diaminobenzidene and chelated with osmium tetroxide to produce osmium black, an electron dense precipitate (7). The immunoperoxidase techniques were popular until the advent of colloidal gold probes.

Faulk and Taylor (14) initiated colloidal gold immunocytochemistry for electron microscopy when they adsorbed antibodies on to colloidal gold spheres and used them to immunolabel antigens on the surface of salmonellae. Romano *et al.* (15) used horse antibody adsorbed to gold spheres for immunolabelling and later described Protein A adsorbed to colloidal gold as an immunological marker (16). Protein A, from the cell walls of staphylococci, binds in a pseudo-immune fashion with certain Ig molecules. Protein G from the cell walls of streptococci reacts similarly and has recently been added to the arsenal of immunological colloidal gold reagents (17, 18) as have streptavidin–gold probes for use with biotinylated antibodies (19).

The production of gold probes depends upon the adsorption of proteins on to colloidal gold. The method was successfully addressed by a report (20) which remains a classical description of the factors affecting adsorption of proteins to colloidal gold. Knowledge of the process has been enhanced by reports of quantification of labelling colloidal gold with Protein A (21) and the immunolabelling efficiency of Protein A–gold complexes (22).

Frens (23) described the preparation of various sizes of colloidal gold, and at present the available size range for electron microscopy is between 1 and 40 nm (24). Homogenous preparations of 5, 10, 15, and 20 nm probes can be distinguished from each other with the electron microscope and are highly suited to multiple immunolabelling experiments.

Colloidal gold probes are ideal for electron microscopy, being particulate and very electron dense. As biological tissue is unlikely to contain a population of homogeneous, electron dense spheres, the colloidal gold probes are distinctive as well as being sufficiently small to permit detailed ultrastructural examination of the immunolabelled specimen. The colloidal gold technique is safe and relatively quick to perform. The probes can be used for both scanning and transmission electron microscopy and their particulate nature is ideal for quantification. Although the use of colloidal gold probes for electron microscopy was first reported in 1971 (14), applications were few until Roth *et al.* (25) described the post-embedding technique. This appears to have been the turning point and there has since been a plethora of new techniques and applications, and colloidal gold probes are almost without exception the first and only choice for electron microscopy.

4. Aims of the book

The aims of this book are to present practical details of the various immunocytochemical techniques in routine use so that the reader may identify the suitability of each technique for a particular application, recognizing its advantages and limitations, and placing it in relation to other immunocytochemical techniques. The next chapter therefore introduces the concepts of immunocytochemistry and the commonly used reagents. Techniques for light (Chapter 3) and electron microscopy (Chapter 4) are described and followed by a chapter devoted to the problems of specimen preparation associated with botanical specimens (Chapter 5). Additional chapters present multiple immunolabelling techniques (Chapter 6) and techniques for image analysis (Chapter 7). A following chapter introduces the highly specialized imaging techniques of video-enhanced light microscopy, high voltage transmission electron miscrocopy, and field emission scanning electron microscopy for the localization of gold particles (Chapter 8). Immunocytochemistry has become so successful that many excellent reagents are commercually available. Gold probes are, however, relatively easy to prepare and for this reason the small-scale laboratory production of these reagents is also described in Chapter 8. Whilst immunocytochemistry is not a central theme for *in situ* hybridization techniques, immunocytochemical methods are involved and a description of the *in situ* hybridization technique can therefore be justified in this book (Chapter 9). In addition, the chapter on trouble-shooting will hopefully help short-circuit some of the tedious teething problems sometimes encountered with immunocytochemistry (Chapter 10). Overall therefore, it is hoped that the book will be a practical guide to the immunocytochemical techniques in use at the present time.

References

1. Hearle, J. W. S. (1972). In *The Use of the Scanning Electron Microscope*, (ed. J. W. S. Hearle, J. T. Sparrow, and P. M. Cross), pp. 1–23. Pergamon, Oxford.
2. Hayat, M. A. (1989). *Principles and Techniques of Electron Microscopy*, p. P1. Macmillan, Hong Kong.
3. Coons, A. H., Creech, H. L., and Jones, R. N. (1941). *Proc. Soc. Exp. Biol. Med.*, **47**, 200.
4. Nakane, P. K. and Pierce, G. B. (1966). *J. Histochem. Cytochem.*, **14**, 929.
5. Avrameas, S. and Uriel, J. (1966). *C.R. Acad. Sci., Paris, Ser. D.*, **262**, 2543.
6. Nakane, P. K. (1968). *J. Histochem. Cytochem.*, **16**, 557.
7. Sternberger, L. A. (1979). *Immunocytochemistry*, pp. 130–69. Wiley, New York.
8. Mason, D. Y. and Sammons, R. E. (1976). *J. Clin. Pathol.*, **31**, 454.
9. Bondi, A., Chieregatti, G., Eusebi, V., Fulcheri, E., and Bussolati, G. (1982). *Histochemistry*, **76**, 153.
10. Campbell, G. T. and Bhatnagar, A. S. (1976). *J. Histochem.*, **24**, 448.
11. Roth, J. (1982). *J. Histochem. Cytochem.*, **30**, 691.
12. Holgate, C. S., Jackson, P., Cowen, P. N., and Bird, C. C. (1983). *J. Histochem. Cytochem.*, **31**, 938.
13. Singer, S. J. and Schick, A. F. (1961). *J. Biophys. Biochem. Cytol.*, **9**, 519.
14. Faulk, W. R. and Taylor, G. M. (1971). *Immunochemistry*, **8**, 1081.
15. Romano, E. L., Stolinski, C., and Hugh-Jones, N. C. (1974). *Immunochemistry*, **11**, 521.
16. Romano, E. L. and Romano, M. (1977). *Immunocytochemistry*, **14**, 711.

17. Bendayan, M. (1987). *J. Electron Micros. Technol.*, **6**, 7.
18. Taatjes, D., Ackerstrom, B., Bjorck, L., Carlemalm, E., and Roth, J. (1987). *Eur. J. Cell Biol.*, **45**, 151.
19. Tolson, N. D., Boothroyd, B., and Hopkins, C. R. (1981). *J. Microsc.*, **123**, 215.
20. Geoghegan, W. D. and Ackerman, G. A. (1977). *J. Histochem. Cytochem.*, **25**, 1182.
21. Horisberger, M. and Clerc, M-F. (1985). *Histochemistry*, **82**, 219.
22. Ghitescu, L. and Bendayan, M. (1990). *J. Histochem. Cytochem.*, **38**, 1523.
23. Frens, G. (1973). *Nature: Phys. Sci.*, **241**, 20.
24. Leunissen, J. L. M. and de Mey, J. R. (1989). In *Immuno-gold Labelling*, (ed. A. J. Verkleij and J. L. M. Leunissen), pp. 3–16. CRC Press, Boca Raton.
25. Roth, J., Bendayan, M., and Orci, L. (1978). *J. Histochem. Cytochem.*, **28**, 55.

2

Immunocytochemical avenues

JULIAN E. BEESLEY

1. Introduction

Biological macromolecules such as proteins, carbohydrates, lipids, nucleic acids, and synthetic polypeptides which are capable of evoking antibody production (1) can be correlated with cell structure using immunocytochemistry. Immunocytochemistry relies upon the specific binding of an antibody to the antigen in the tissue, the reaction being localized with respect to cell structure by attaching a microscopically dense marker to the antigen–antibody complex. Immunocytochemistry is highly specific, relatively quick, and relatively sensitive. Many of the reagents are commercially available and the techniques are now routine.

Selection of an immunocytochemical technique implies that the antigen must be stabilized *in situ* and in a reactive form for incubation with the specific antibody and the microscopically dense probe. The antibody, the specimen preparation technique, the microscopically dense marker, the immunolabelling technique, and of course the validity of the result must be considered before immunolabelling to achieve a high signal to background ratio for unequivocal results. These considerations, which are reviewed here as an introduction to the following chapters, are identical for both histological and electron micropscope immunocytochemistry.

2. The antibody

Antibodies have been described in detail in previous volumes in this series (2, 3), therefore only those facts pertinent to immunocytochemistry will be considered here.

The antibody is central to the success of immunocytochemical labelling. A good antibody is highly specific, possessing a high titre (amount of antibody activity (1)), high affinity (the exactitude of stereochemical fit of an antibody-combining site to its complementary antigenic determinant (2)), and high avidity (the total binding stength of a multivalent antigen with the antibody (4)). Many antibodies are available commercially. Both polyclonal and monoclonal antibodies can be used for immunocytochemistry, the choice being entirely dependent on antibody performance in immunocytochemical marking. An antibody molecule is Y-shaped, the two upwardly pointing arms being the antibody-binding sites or Fab fragments. Cleavage of the antibody molecule with papain separates the Fab fragments which can still bind to antigen and can be used for high resolution immunolabelling (5). Since immunocytochemical reagents are linked in a chain from antigen to microscopically dense marker, it is important to be absolutely certain of the class and sub-class of the antibody since this will determine other reagents used.

It is important that the antibody is thoroughly characterized before use in an experimental situation. Characterization can be by non-microscopical means such as immunoblotting, but ideally the antibody should be tested by immunocytochemistry on a known positive tissue because a good antibody for one purpose may not always be ideal for another.

Polyclonal and monoclonal antisera each possess distinctive characteristics useful for a range of immunocytochemical techniques. Immunocytochemistry must therefore appreciate and capitalize on the characteristics of each type of antiserum.

2.1 Polyclonal sera

Polyclonal sera contain a mixture of high affinity antibodies, each active against different epitopes on the antigen, and provided that not all these epitopes are destroyed by fixation, immunolabelling remains possible. For electron microscope studies the antigen can usually tolerate glutaraldehyde fixation, thereby achieving reasonable ultrastructure. Production of polyclonal antisera is relatively simple and of low cost. In addition polyclonal sera possess good stability at varying pH and salt concentrations, there is no risk of loss of clones, and they are useful for pre-adsorption controls (4).

Polyclonal antisera are produced by injecting an antigen into an host animal, usually either rabbit, sheep, or goat and, after booster injections, the serum is periodically collected and tested for antibody. Antibodies, which are glycoproteins, are produced in the spleen of the host animal by B lymphocytes and plasma cells, the latter being the terminal stage in the differentiation of B lymphocytes. Sites on the complex antigenic molecule that are recognized by antibody are called epitopes (4). A polyclonal antiserum is a mixture of antibodies to different epitopes on the immunogen. The antiserum will probably contain antibody to impurities in the immunogen. Antibodies raised against the contaminating immunogens are often of low titre and affinity and can be diluted out to zero activity for immunolabelling. In consequence of the high possibility of a wide spectrum of antibodies being present in the host animal in response to previous antigen challenges, serum removed from the animal before injection of the immunogen is important as a negative, or pre-immune, control (Section 6 below).

2.2 Monoclonal sera

A monoclonal serum is a 'pure' preparation of one of the constituents of polyclonal antiserum. Monoclonal antisera are useful as diagnostic reagents and for immunocytochemical research since large amounts of antibody with narrow specificity, titre, and homogeneity in all batches can be produced. A low antigen dose is needed for immunoresponse and the antigen need not be purified with as much stringency as for the production of polyclonal antisera. It is indeed possible through careful screening of the hybridomas to find antibodies raised against rare or unidentified antigens. The activity of a monoclonal antibody is, however, susceptible to fixation of the antigen since monoclonal antibodies are specific to a single epitope on the antigen and if this is destroyed all reactivity is lost. For electron microscopy, relatively weak fixation, as achieved with formaldehyde, is usually necessary and in consequence ultrastructure is poor. Formaldehyde is the routine fixative for histological studies and specimen morphology is therefore not compromised.

The initial phase of preparation is to induce synthesis of a polyclonal antiserum in the host animal as described above. The plasma cells are removed from the host animal and each fused with a malignant myeloma cell to form hybridoma cells (6). Hybridoma cells are isolated and cultured and theoretically each will produce unlimited monoclonal antibodies raised against a specific epitope, each antibody being identical in reactivity and titre. These hybridoma cells are either cultured *in vitro*, in which the tissue culture supernatant fluid contains antibody, or *in vivo* in the peritoneal cavity of a suitable host and it is the ascites fluid which contains antibody. The latter is a convenient method of producing large amounts of antibody, 10^2 to 10^3 times more concentrated than tissue culture supernatant, but the serum is no longer monospecific since the ascites contains antibodies from the host animal (4). The serum can, however, be purified.

3. Specimen preparation

Identical criteria should attend specimen preparation for immunocytochemistry as for other microscopical studies—the morphology of the specimen must be commensurate with the resolution of microscopic analysis. An additional requisite for immunocytochemistry is that specimen preparation should maintain the status quo of the epitope, by retaining its capacity to bind antibody, by retaining its position within the cell, and by ensuring that the epitope is in a configuration on the specimen such that antibodies can physically interact with it. Light and electron microscope studies are frequently complementary. Histological investigations give an overall view of immunolabelling, identifying the number of positive cells, whereas electron microscopy identifies immunolabelling associated with the individual cells.

The specimen can be used fixed or unfixed. Fixation denatures the proteins and stabilizes the cells against dehydration, embedding, sectioning, and staining. Cross-linking fixatives such as *p*-benzoquinone, diethylpyrocarbonate, and the aldehydes achieve their effects through formation of cross-links between reactive groups in the polypeptide chains of proteins, whereas the dehydrating fixatives—the alcohols—disrupt the H–H bonds in proteins. Fixation may severely reduce the potential of an antigen to bind antibody. Fixation should be selected carefully to preserve sufficient morphology commensurate with the resolution required whilst still retaining antigenic activity.

If unfixed whole cells are immunolabelled capping may occur. The first step in capping is the formation of antigen–antibody complexes into patches as a result of cross-linking by bivalent or multivalent antibodies. These patches flow to and accumulate at one pole of the cell where they form a large mass, the cap, which may be internalized. Unlike formation of patches, capping is an energy-dependent process that may require microfilaments attached to the cytoplasmic side of the membrane. Although capping has been extensively studied in lymphocytes it may also take place in many other cells (1). Polyclonal sera contain antibodies to many epitopes on each antigen and immunolabelling is generally dense. The possibility of cross-linking ligands with the secondary reagent thereby inducing capping, is high. Minimal capping occurs during the use of monoclonal antibodies since these are directed against a single epitope on each antigen and immunolabelling is relatively sparse. Capping can be prevented by using a monovalent ligand such as the Fab fragment of an antibody.

Histological immunocytochemistry is relatively rapid compared with electron microscope studies. Histological preparations are routinely either smears or tissue sections from wax-embedded or frozen blocks of tissue. Histological immunocytochemistry is usually less problematical than electron microscopical immunocytochemistry because the constraints of resolution are much less and mild fixatives such as neutral buffered formalin or methanol, which inflict minimal damage to the antigen, can be used. It is advisable, if possible, to perform light microscope immunocytochemistry before attempting electron immunocytochemistry.

For electron microscopy, the structure–function compromise is acute. Electron microscopy is a high resolution morphological technique and any compromise on fixation, to preserve antigenicity, is noticeable. Routine fixatives are formulations of either glutaraldehyde and/or formaldehyde depending upon the antibody being used (Section 2). The pre-embedding, the immunonegative stain, and the scanning electron microscope techniques are used routinely to obtain antigenic information from the outer surface of the plasma membrane, whereas the post-embedding and occasionally the pre-embedding techniques are used to investigate intracellular antigens.

4. The microscopically dense marker

The microscopically dense marker enables the site of the antigen–antibody reaction to be correlated with the morphology of the specimen. The marker should highlight the antigen–antibody reaction without obliterating the underlying structure of the specimen, without apparent displacement of the antigen–antibody complex, and should be specific for the ligand in question. The marker contains two components, a microscopically dense reagent and a reagent which binds to the primary antibody. The microscopically dense reagent possesses either intrinsic density or potential density which can be induced chemically. For light microscopy, it could be an enzyme, such as peroxidase or alkaline phosphatase, which is developed to a coloured reaction product. The fluorescent techniques remain popular. The use of 5 nm colloidal gold probes, silver enhanced to yield a dense black reaction product, are also used by many workers. For electron microscopy, colloidal gold probes produce superior resolution to the enzyme techniques and are the only probes in current routine use. Conjugation of a binding protein to the dense reagent ensures linkage of the dense marker to the primary antibody. The binding protein can be an antibody against the Ig of the primary antibody species or a protein, such as Protein A or Protein G, which reacts specifically with certain antibody molecules. If the primary antibody has been biotinylated the marker will contain streptavidin. Microscopically dense markers for application with a wide variety of primary antibodies are commercially available.

5. Immunolabelling techniques

The antigen, the antibody, and the microscopically dense probe must be linked together by successive incubations. It is essential therefore that all the reagents are compatible.

The most common immunolabelling technique is the indirect, or two step, technique in which the antigen is incubated with specific antibody followed by the dense marker. The indirect technique is widely used because a single dense marker can be used to localize a number of different unlabelled primary antibodies, each raised against different antigens but each with a similar affinity for the dense marker.

There may be instances when a dense marker of the preferred specificity is not available. A secondary antibody can be interspersed between the primary antibody and the dense marker. This three step technique is usually the longest technique attempted although the chain length could be longer. The peroxidase–anti-peroxidase technique (7) is a specialized, highly sensitive three step technique. The second or bridging antibody is applied in excess so that one antibody binding site remains free to bind the microscopically dense

11

probe which is an enzyme–antibody complex, the antibody being from the same species as the primary host species.

If there is an abundance of primary antibody and the same immunocytochemical test is to be performed routinely, the dense marker can be coupled directly with the primary antibody. Immunolabelling is therefore effected in a one-step or direct technique.

These immunolabelling methods are discussed in detail in Chapter 3.

All incubations are usually for 1 h and are performed at room temperature. It will be assumed that if any unknown reagents are used a series of dilutions of reagents in a suitable buffer will be tested to optimize immunolabelling.

Non-immunological attachment of antibodies and dense markers to the specimen can be reduced by reagents added to the buffers. Some of these reagents are applied during pre-incubations while others are included in the buffers used for washing the specimen and for diluting the antibodies and dense markers. These reagents are of three types:

(a) Blocking reagents, such as bovine serum albumin and gelatin, are proteins which compete with the immunological reagents for non-immunological sticky sites on the specimen.

(b) Surfactant reagents, such as Tween 20 and NaCl, reduce surface charges on the specimen, thereby reducing electrostatic attachment of molecules to the specimen.

(c) Chemical reagents, such as 0.02 M glycine, or 1% sodium borohydride block free aldehyde groups on the tissue thereby preventing fixation of proteins to the specimen.

The buffer most often used is phosphate (0.01 M, pH 7.2)-buffered saline (0.15 M), containing 1% bovine serum albumin (8).

Excess immunological reagents are removed by washing with buffer. After immunolabelling it is customary, although not always necessary, to stabilize the immunological reaction by fixation. A final wash in water removes the buffer and the specimen is contrasted prior to examination.

6. Validation of the result

It is possible to achieve 'credible' non-specific labelling on a specimen. Likewise the absence of immunolabelling does not always signify the absence of antigen. Controls are therefore absolutely essential and should be devised specifically to verify each experiment. Immunological controls should be performed before immunolabelling in a test system to determine the exact specificity of the antibody. Methodological controls assure the quality of the immunolabelling technique.

Positive controls test the reactivity of the antigen, the antibody, and the dense marker. For instance, if possible the antigen in the specimen should be immunolabelled with an alternative antibody to prove that specimen preparation procedures have not destroyed the antigen. Similarly the efficacy of the test antibody should be confirmed by immunolabelling the antigen in a different specimen. Finally the dense marker should be tested in a known positive system.

Negative controls assess the methodological non-specificity of the immuno-labelling technique employed. Methodological non-specificity can be assessed by immunolabelling a specimen where the antigen is known to be absent or immunolabelling the test specimen with a pre-immune or negative serum. Additionally, the antibody should be pre-adsorbed with excess antigen and used for immunolabelling the test specimen.

7. Conclusions

All immunolabelling is in principle identical, and all immunolabelling techniques can be reduced to a number of constitutive steps. Once it has been decided that immunocytochemistry will be used in a study, consideration of the antibody, the specimen preparation technique, and the microscopical dense marker will indicate an appropriate immunolabelling technique to employ, and if approached with care and with sufficient control incubations there is no reason to suppose that anything but success will follow.

References

1. Rosen, F. S., Steiner, L. A., and Unanue, E. R. (1989). *Dictionary of Immunology*. Macmillan, London.
2. Catty, D. (ed.) (1988). *Antibodies: A Practical Approach*, Vol. I. IRL Press, Oxford.
3. Catty, D. (ed.) (1989). *Antibodies: A Practical Approach*, Vol. II. IRL Press, Oxford.
4. Beltz, B. S. and Burd, G. D. (1989). *Immunocytochemical Techniques, Principles and Practice*. Blackwell Scientific, Massachusetts.
5. Baschong, W. and Wrigley, N. G. (1990). *J. Electron Microsc. Tech.*, **14**, 313.
6. Kohler, G. and Milstein, C. (1975). *Nature*, **256**, 495.
7. Sternberger, L. A. (1979). In *Immunocytochemistry*, pp. 108–15. Wiley, New York.
8. Slot, J. W. and Geuze, H. J. (1984). In *Immunolabelling for Electron Microscopy*, (ed. J. M. Polak and I. M. Varndell), p. 129. Elsevier, Amsterdam.

3

Immunolabelling techniques for light microscopy

PETER JACKSON and DAVID BLYTHE

1. Introduction

Immunofluorescence (1) remains a widely used technique for the identification of immunoglobulins and complement in skin and renal biopsies. The success of immunofluorescence in some areas of pathology stimulated interest in the development of alternative antibody-labelling techniques which would avoid the difficulties and limitations associated with immunofluorescence. Many limitations were overcome with the introduction of enzymes as antibody labels (2, 3). Recent progress in immunocytochemistry has resulted in the development of immunocytochemical methods that, despite their increasing sophistication, are sufficiently reproducible and easy to perform and which are no longer the exclusive tool of the researcher. All modifications of immunocytochemical techniques have been primarily developed to give increased specificity and amplification to the reaction. This chapter provides detailed laboratory immunolabelling protocols for techniques which may be applied to the full range of histological preparations.

1.1 Definition

Immunocytochemistry can be considered as the demonstration of antigens in tissue sections or smears by the use of specific antigen–antibody interactions which culminate in the attachment of a marker to the antigen. For light microscopy, the marker may be a fluorescent dye, an enzyme, or colloidal gold. The aim is to achieve reproducible and consistent demonstration of specific antigens with the minimum of background staining whilst ideally, preserving the integrity of tissue architecture.

2. Immunolabelling methods

There are numerous immunocytochemical techniques which may be used to localize antigens. The selection of a suitable technique should be based on parameters such as the type of specimen under investigation and the degree of sensitivity required. Cost may also be a factor.

2.1 Direct method

In this technique a labelled antibody reacts directly with the antigen in the histological or cytological preparation (*Figure 1*). This method utilizes only one antibody and can be quickly completed. The technique, however, provides little signal amplification and is rarely used on paraffin sections. The main application remains in immunofluorescence, where this technique is used to identify immunoglobulin and complement in frozen sections of skin and renal biopsies.

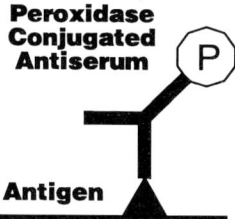

Figure 1. Direct immunolabelling method (see text for description of immunolabelling sequence).

2.2 Two step indirect method

An unlabelled primary (first layer) antibody is visualized by a labelled secondary (second layer) antibody directed against the immunoglobulin of the animal species in which the primary antibody has been raised (*Figure 2*).

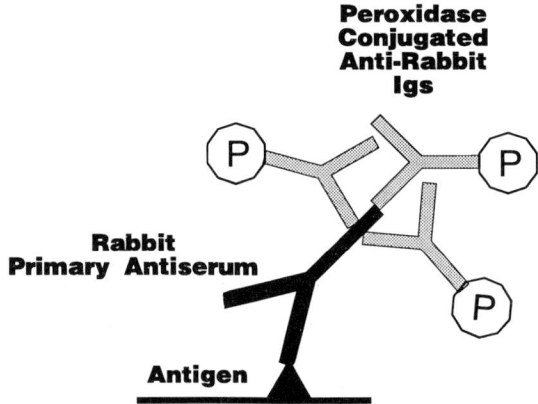

Figure 2. Two step indirect immunolabelling method (see text for description of immunolabelling sequence).

This method is more sensitive than the direct method because several secondary antibodies may react with different antigenic sites on the primary antibody. Consequently more labelled secondary antibody binds to the antigenic sites. The technique is more versatile than the direct method as the same labelled secondary antibody can be used with a variety of primary antibodies raised from the same animal species.

2.3 Protein A method

Protein A is derived from the cell wall of the bacterium *Staphylococcus aureus*, and binds to the Fc portion of IgG molecules from several mammalian species (4). Protein A may be used as a secondary reagent for the demonstration of antigens by light microscopy when labelled with suitable enzymes or colloidal gold (*Figure 3*).

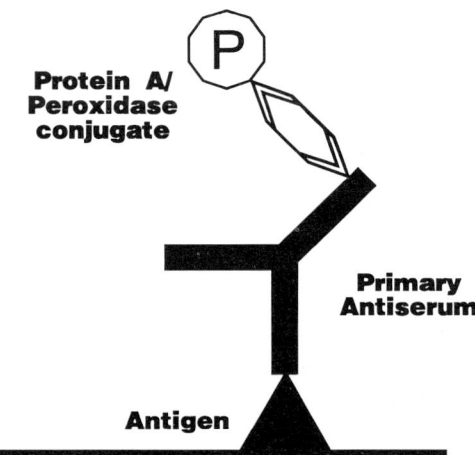

Figure 3. Protein A immunolabelling method (see text for description of immuno-labelling sequence).

2.4 Unlabelled antibody methods

These indirect methods utilize a pre-formed, cyclic, enzyme–anti-enzyme immune complex composed of three enzyme molecules and two antibody molecules. The methods are named after the particular enzyme–antibody complex that is used in the technique. The peroxidase–anti-peroxidase (PAP) technique (*Figure 4*) utilizes a peroxidase–anti-peroxidase immune complex. The alkaline phosphatase–anti-alkaline phosphatase (APAAP) technique (*Figure 5*) uses a similar enzyme–anti-enzyme immune complex consisting of alkaline phosphatase and anti-alkaline phosphatase molecules. Such methods are amongst the most sensitive immunocytochemical techniques and give excellent results on fixed, paraffin-processed material and can be applied equally well on cryostat sections and smears.

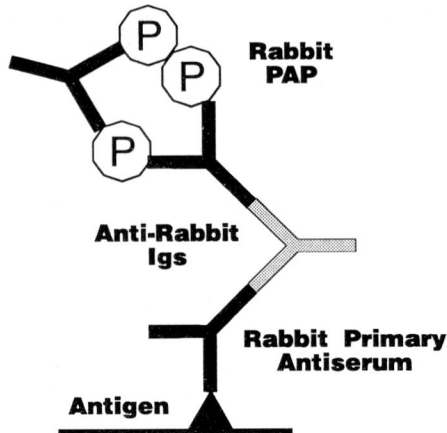

Figure 4. Peroxidase–anti-peroxidase immunolabelling method (see text for description of immunolabelling sequence).

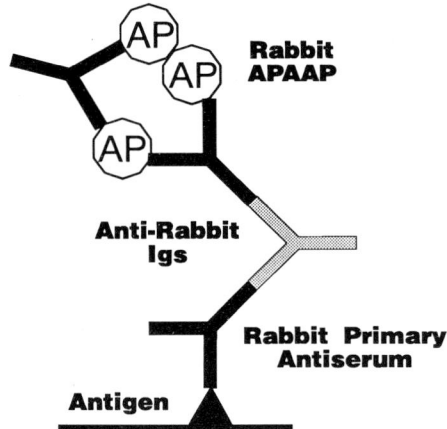

Figure 5. Alkaline phosphatase–anti-alkaline phosphatase immunolabelling method (see text for description of immunolabelling sequence).

2.5 Avidin–biotin methods

Avidin is a basic glycoprotein (mol. wt. 68 kDa) which has a high affinity for the small (mol. wt. 244 Da) water-soluble vitamin biotin. Recently methods have been devised to exploit the high efficiency and specifity of the avidin–biotin reaction. Biotin can be conjugated to a variety of biological molecules, including antibodies. Many biotin molecules can be attached to a single molecule of protein. The biotinylated protein can thus bind to more than one molecule of avidin. However, avidin has two distinct disadvantages when used in immunocytochemical detection systems.

It has a high isoelectric point of approximately 10 and is therefore positively charged at neutral pH. Consequently it may bind non-specifically to negatively charged structures such as the nucleus. The second disadvantage is that avidin is a glycoprotein and reacts with molecules such as lectins via the carbohydrate moiety.

These two problems are overcome with the substitution of streptavidin for avidin. Streptavidin is a protein (mol. wt. 60 kDa) isolated from the bacterium *Streptomyces avidinii*, and like avidin, has four high affinity binding sites for biotin. Streptavidin has an isoelectric point close to neutral pH and therefore possess few strongly charged groups at the near neutral pH used in immunocytochemical detection systems. Furthermore streptavidin is not a glycoprotein and therefore does not bind to lectins. The physical properties of streptavidin therefore make this protein much more desirable for use in immunocytochemical detection systems than avidin.

The most sophisticated and sensitive extension of the streptavidin–biotin technique employs pre-formed complexes. Streptavidin and biotinylated enzyme are simply mixed at appropriate concentrations and allowed to stand for at least 30 min at room temperature for the complex to form. This pre-formed complex is attached to the biotinylated antibody. Careful stoichiometric control ensures that some binding sites remain free to bind with biotinylated antibody. This allows the pre-formed complex to bind and provides a very high signal at the antigen binding site (*Figure 6*).

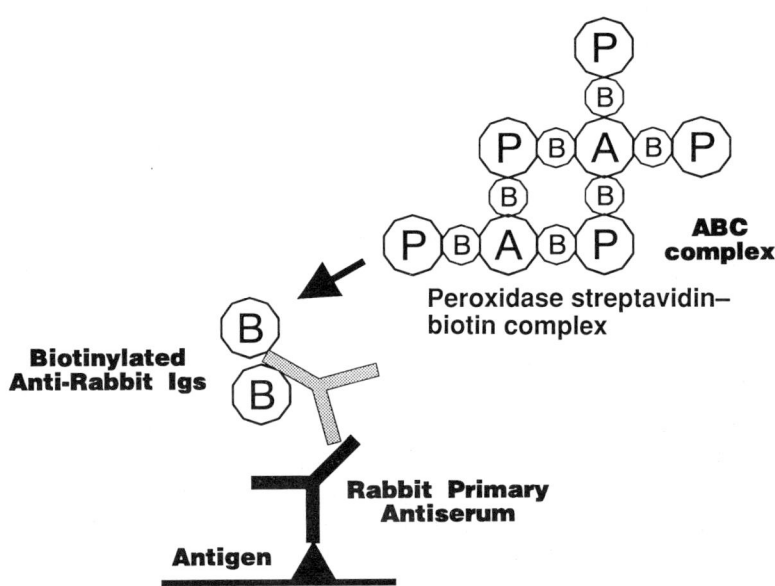

Figure 6. Avidin–biotin complex immunolabelling method (see text for description of immunolabelling sequence).

2.6 Immunogold methods

Immunogold reagents consist of immunological reagents adsorbed on to the surface of colloidal gold particles. Detailed methods of complexing colloidal gold particles with immunoglobulins and other macromolecules are reviewed by Roth (5). Colloidal gold-labelled antibodies, introduced by Faulk and Taylor (6), were limited to immunocytochemical studies for electron microscopy. Geoghegan *et al.* (7) demonstrated surface immunoglobulins on B cells for both electron microscopy and bright field light microscopy using a two layer indirect technique employing a gold-labelled secondary antibody. In general, however, such techniques are not acceptable for bright field light microscopy as the technique requires relatively high concentrations of gold-labelled antibody, long incubation times and the use of high magnification objectives. In light microscopial studies immunogold-labelled sites appear red. The best results have been shown by antibodies adsorbed onto 20 nm gold particles (8).

Considerable amplification of the signal can be obtained by silver enhancement of the colloidal gold particles. This is best achieved by the use of a physical developing solution. Danscher (9), described a physical developing solution containing silver lactate for the detection of metallic gold in tissue sections. In 1983, Holgate *et al.* (10) used a modification of Danscher's physical developer to localize gold-labelled immunoglobulins in paraffin sections of lymphoid tissues. This method was named the immunogold–silver staining method (*Figure 7*).

Silver enhancement of colloidal gold uses the property of metallic gold to catalyse the reduction of silver ions to metallic silver in the presence of a reducing agent. During this reaction the gold particles become coated in metallic silver, which eventually becomes visible when viewed with the bright field microscope.

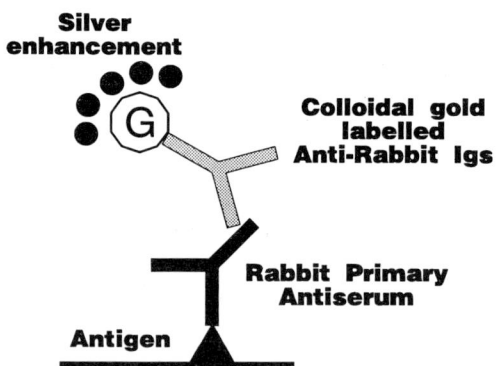

Figure 7. Immunogold–silver staining method (see text for description of immuno-labelling sequence).

Immunogold–silver staining produces a signal that is insoluble in de-hydrating and clearing agents and is visible with low power light microscopy. Endogenous pigments cannot be confused with the intense reaction product and endogenous enzymes are not developed as in the immunoenzymatic techniques and therefore do not interfere with the interpretation. There are no carcinogenic chromogens associated with the technique, and furthermore, the reaction product may be enhanced by epi-polarization microscopy (11) or dark field observation (12).

The best immunogold reagents for bright field microscopy immunocyto-chemistry with silver enhancement are those labelled with 1 nm or 5 nm gold particles. Small gold particles allow efficient tissue penetration in cryostat, paraffin-, and resin-embedded sections. The small gold particles also suffer minimal steric hindrance thereby allowing higher labelling intensity at the antigenic sites.

3. Chromogens

Except for the colloidal gold methods, immunocytochemical staining methods for light microscopy depend upon enzyme–substrate reactions which convert colourless chromogens into visible, coloured end products. Since the introduction of horseradish peroxidase as an antibody label, reliance has been placed on the sensitive hydrogen peroxide–diaminobenzidene reaction, which produces a brown end product insoluble in alcohol, xylene, and other inorganic solvents.

The diaminobenzidene (DAB) reaction product can be intensified by the addition of Imidazole (1 mg/ml) to the substrate solution, or by subsequent incubation of the sections in a copper sulphate solution (see the Appendix). Furthermore, the DAB reaction product has the ability to react with osmium tetroxide, increasing its staining intensity for electron microscopy. With regard to the possible carcinogenic properties of DAB care should be taken when handling the dry powder. To avoid repeated handling we suggest that the DAB solution should be prepared in bulk, aliquoted and stored frozen (see the Appendix).

Publicity regarding the possible carcinogenic properties of DAB has led to the use of alternative chromogens for peroxidase development. These substances have generally been found to be less satisfactory than DAB, as the coloured end products are soluble in laboratory dehydrating agents or xylene-based mounting media. Such chromogens include 3-amino-9-ethylcarbazole (AEC) (*Protocol 12*) and 4-chloro-1-naphthol (CN) (*Protocol 13*). These substances produce a red and a dark blue reaction product respectively, and are therefore useful in double labelling techniques. AEC-stained sections may fade in excessive light and should be stored in the dark. The CN reaction product tends to diffuse on storage.

4. Fixation

A prerequisite for all histological and cytological investigations is to ensure preservation of tissue architecture and cell morphology by adequate and appropriate fixation. Furthermore, an ideal fixative should stabilize and protect tissues and cells from the damaging effects associated with subsequent treatment. Reagents used for this purpose may significantly diminish the antibody binding capability. The demonstration of many antigens depends heavily on the fixative employed and the immunocytochemical method selected. Prompt fixation is essential to achieve consistent results. Poor fixation, or delay in fixation, causes loss of antigenicity or diffusion of antigens into the surrounding tissue. There is no one fixative that is ideal for the demonstration of all antigens and many antigens necessitate the use of frozen sections. However, in general, many antigens can be successfully demonstrated in formalin-fixed paraffin-embedded material. The most widely used fixatives in diagnostic hospital histology laboratories are formalin based; neutral buffered formalin, formal saline, or 10% formalin in tap water being the three most commonly employed for general use. In cases where antigen demonstration is poor, improved immunocytochemical staining may be achieved by the use of specially formulated fixatives. Excellent cytoplasmic immunoglobulin demonstration can be obtained by using a mercuric chloride-based fixative such as B5 (see the Appendix). This solution is highly recommended for the fixation of lymph node biopsies. Surface membrane immunoglobulin, however, is not readily demonstrated following B5 fixation.

Periodate–lysine–paraformaldehyde (PLP) (see the Appendix), is a fixative designed to protect the carbohydrate moiety associated with cell membranes against the damaging effects of tissue processing and embedding (13). The periodate oxidizes the carbohydrate to produce aldehydes which are cross-linked with lysine, whilst the paraformaldehyde gives good preservation of the ultrastructure. Although, presumably, designed as a fixative for electron microscopy PLP may be used for light microscopy giving better demonstration of some surface membrane-associated antigens than with formalin fixation. The tissue morphology, however, is usually poorer on paraffin-embedded material than that associated with formalin fixation.

Holgate *et al.* (14) modified PLP by the addition of 5% potassium dichromate solution (a lipid fixative). This novel fixative was abbreviated to PLP/D (see the Appendix). This modification was an attempt to ensure that both the carbohydrate and lipid moeity associated with cell membranes were stabilized against further denaturation and/or mobilization during tissue processing. The integrity of the cell membrane and therefore the antigenic determinants are better preserved than with the PLP fixation alone. Protection of antigenicity is, however, not complete as more labile tissue antigens such as those associated with T-cell subsets, although demonstrable in frozen sections of PLP/D-fixed tissue, cannot be demonstrated in tissues processed to paraffin wax. Furthermore the overall morphological preservation is not ideal using this fixative.

In summary good tissue fixation and hence morphology, together with good antigen demonstration do not necessarily concur and a compromise may have to be achieved.

5. Histological section preparation techniques

5.1 Paraffin sections

Since its introduction, paraffin wax has remained the most widely used embedding medium for diagnostic histopathology in routine histological laboratories. Accordingly, the largest proportion of material for immunocyto-chemistry is formalin-fixed, paraffin-embedded (*Figure 8*). Formalin-fixed, paraffin-embedded material produces satisfactory results for the demonstration of the majority of tissue antigens. The demonstration of many antigens can be significantly improved by treatment with proteolytic enzymes such as trypsin

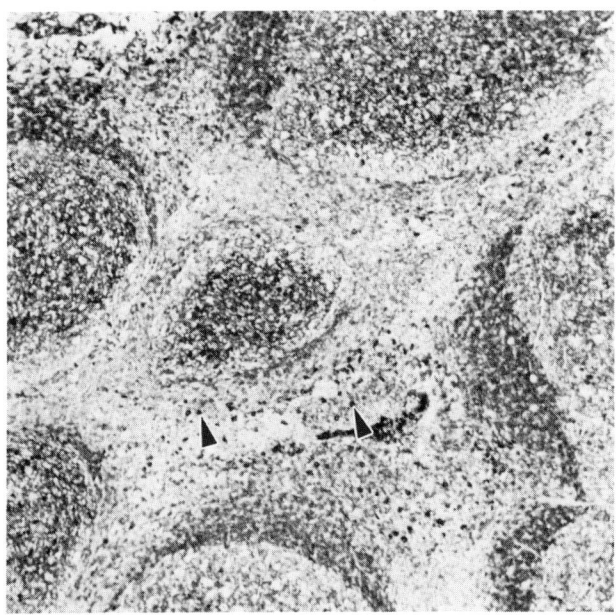

Figure 8. Paraffin section of normal tonsil prepared as described in Section 5.1 demonstrating kappa light chains in plasma cells (arrows) immunolabelled using the streptavidin–biotin technique (*Protocol 4*). A change in the distribution of the kappa light chains would indicate the presence of a malignant tumour.

(15). Antigen demonstration is improved due to the enzyme breaking the protein cross-links formed by formalin fixation and thereby exposing hidden antigenic sites. Not all antigens require proteolytic digestion and care must be taken not to create 'false' antigenic sites or to alter existing antigenic sites. In some instances immunostaining is reduced, or absent, following enzyme digestion. Individual laboratories should investigate which antibodies require proteolytic digestion, using their own fixed, paraffin-processed material.

For general use trypsin is the most widely used proteolytic enzyme, although for renal immunocytochemistry we find that protease gives optimum antigen demonstration. Other proteolytic enzymes which can be employed include chymotrypsin, pepsin and proteinase K.

For optimal immunolabelling, trypsinization time is critical, and depends upon the length of time of fixation. Tissues fixed in formalin for long periods usually require prolonged trypsinization. For extended trypsinization times, the trypsin solution should be changed after 30 min. The digestion time will also depend upon the type of trypsin used and this may vary with each batch.

Where fixation and paraffin processing schedules are not known, as with referred blocks and slides, it may be necessary to perform a range of digestion times. In these cases it is worth recording the optimum digestion time which may be useful for future immunocytochemical studies on material sent from the same source.

The immunocytochemical methods listed in this chapter have been adapted for material fixed in 10% neutral buffered formalin for 18–24 h, followed by a 16 h automatic tissue-processing schedule. Dehydration is via graded ethanols, clearing by chloroform, and impregnation and embedding in Ralwax 1 (BDH, UK). Sections are cut at 4 μm, picked up on aminopropyltri-ethoxysilane (APES) coated slides (see the Appendix). Sections are allowed to drain before hotplating at 60 °C, for 30 min. Some laboratories prefer to dry sections at 37 °C overnight; however, we achieve good immunostaining using hotplated sections. One exception to this has been found to be in the demonstration of proliferating cell nuclear antigen (PCNA). Such sections need to be dried at 37 °C, otherwise poor immunostaining is obtained.

5.2 Frozen sections

Certain cell membrane antigens (such as T-cell sub-set antigens) do not survive routine fixation and wax embedding. Such antigens can be successfully demonstrated only on frozen sections. Although there are an increasing number of antibodies available that will demonstrate cell membrane-associated antigens in paraffin sections, the use of frozen sections still remains essential for the demonstration of many antigens. The demonstration of T- and B-cell membrane-associated antigens on frozen sections is essential in the diagnosis of lymphoreticular disease, and the use of selected panels of antibodies allow accurate immunophenotyping.

Frozen sections have certain inherent disadvantages when compared to their paraffin counterparts. These include:

- poor morphology
- poor resolution at higher magnifications
- availability of fresh tissue. Special arrangements are required for the collection and storage of fresh unfixed tissues
- only limited retrospective studies may be carried out
- hazards to health associated with the handling of fresh unfixed material, e.g. HIV, hepatitis B, and TB
- frozen sections are technically more difficult to cut

Tissue should be received as soon after removal as possible. Good liaison with clinicians and theatre staff is essential. In cases when delay is inevitable, we find that specimens should be transported in tissue culture medium, such as RPMI, and cooled by attaching the specimen container to an ice pack.

Protocol 1. Preparation of frozen sections for immunolabelling

1. Snap freeze small pieces of tissue (5 × 3 × 3 mm) using pre-cooled isopentane in liquid nitrogen.
2. Cut frozen sections at 5 μm. Pick up sections on APES-coated slides.
3. Air-dry the sections overnight at room temperature.
4. Fix sections in acetone for 20 min.
5. Air-dry sections.
6. Proceed with immunostaining (see protocols) or store sections at −20 °C. Sections stored at −20 °C should be wrapped in aluminium foil for protection until required.

Acetone fixation does not completely stabilize frozen sections against the detrimental effects of long incubations in aqueous solutions. Following long immunocytochemical techniques, acetone-fixed frozen sections may show morphological changes such as chromatolysis and apparent loss of nuclear membranes. We have found that such changes to tissue morphology can be prevented by ensuring that the sections are thoroughly dried both before and after fixation in acetone.

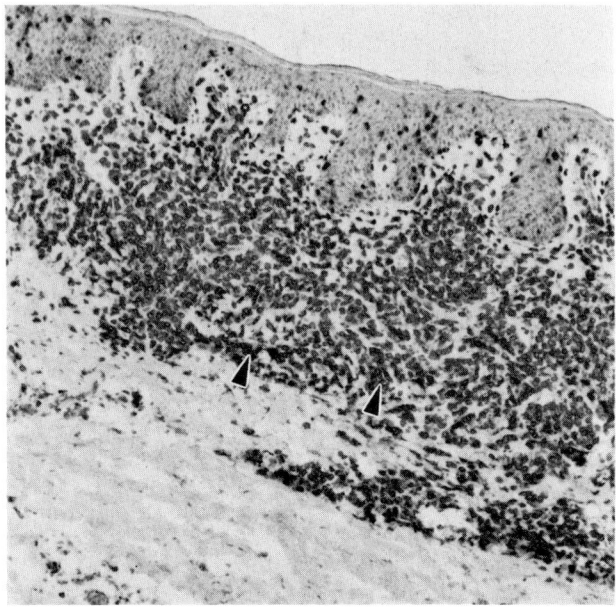

Figure 9. Frozen section of skin prepared as in *Protocol 1* stained with the anti-CD3 antibody for the demonstration of T cells (arrows) using the streptavidin–biotin technique (*Protocol 5*).

5.3 Resin sections

Embedding tissues in resin offers several advantages over frozen or paraffin sections:

- improved morphology
- less tissue shrinkage
- calcified material may be sectioned without prior decalcification
- infiltration of acrylic resins and polymerization can take place at low temperatures
- thinner sections may be cut

The two groups of resin most commonly used in histology are epoxy and acrylic resins.

5.3.1 Epoxy resins

The various epoxy resins have found their widest application as embedding media for ultrastructural studies. A few immunocytochemical studies for light microscopy have been reported, but these techniques rely on etching the sections prior to staining, often with aggressive agents and conditions such as sodium ethoxide (or methoxide) at 60 °C. However, this is unsuitable for immunocytochemical investigations as many tissue and lymphoid antigens fail to demonstrate in epoxy resin sections.

5.3.2 Acrylic resins

Acrylic resins used in histology are esters of acrylic acid or more commonly methacrylic acid. Various methacrylates have been tried, alone or in the form of mixtures, but only three have gained widespread acceptance as suitable embedding media; methylmethacrylate, 2-hydroxyethylmethacrylate (glycol methacrylate), and *n*-butylmethacrylate.

Acrylic resins are able to polymerize at 4 °C or less thus avoiding the use of prolonged heat (approx 60 °C) that is necessary for polymerization of epoxy resins or infiltration of paraffin wax, thereby reducing loss of antigenicity.

Glycol methacrylate resins such as JB4 (Polysciences Inc. USA), Immunobed (Polysciences Inc. USA), and Historesin (Leica, UK) are the most commonly used resins for light microscopy. Unfortunately, good immunocytochemical staining is often difficult to achieve using these resins, partly because polymerized glycol methacrylate cannot be removed from tissue sections and hence can form a physical barrier to subsequent immunocytochemical reagents.

We have achieved good immunostaining, however, using a modified methyl methacrylate resin developed by Hand (16). This resin, which is soluble, consists of methylmethacrylate with dibutylphthalate as the plasticizer, *N,N'*-dimethylaniline as an accelerator, and benzoyl peroxide as catalyst. This method in our hands produces comparable immunostaining to that obtained on paraffin sections, but has the added advantage of giving improved morphology and resolution. This has proved extremely useful for bone marrow trephines, where good immunostaining can be achieved on thin undecalcified sections. It also has proved useful in skin, lymph node, and renal immunopathology.

Protocol 2. The preparation of sections of specimens embedded in methylmethacrylate for immunolabelling

1. Fix thin slices of tissue (2 mm thickness) in 10% formalin for 18–24 h.
2. Dehydrate through graded ethanol (50%, 70%, 95%, absolute ethanol).
3. Infiltrate overnight in the monomer mixture:
   - 15 ml methylmethacrylate
   - 5 ml dibutylphthalate
4. Embed, using thick, heavy-duty plastic moulds. Embedding media:
   - 15 ml methylmethacrylate
   - 5 ml dibutylphthalate

Protocol 2. *Continued*

- 1 g dry benzoyl peroxide[a]
- 200 μl *N,N'*-dimethylaniline

5. Polymerize under vacuum for 2 h (minimum).

6. Cut sections at 1 or 2 μm and float out sections on a warm-water bath.

7. Pick up sections on APES-coated slides.

8. Allow sections to drain (do not hotplate, as this diminishes immuno-staining).

9. Dry sections at 37 °C, overnight.

10. Immunostain as for paraffin sections.[b]

[a] Benzoyl peroxide must be completely dry before use, otherwise the embedding solution will turn milky. Care must be taken as benzoyl peroxide is potentially explosive when dry.
[b] The resin is removed prior to immunostaining by treatment with xylene at 37 °C for 30 min.

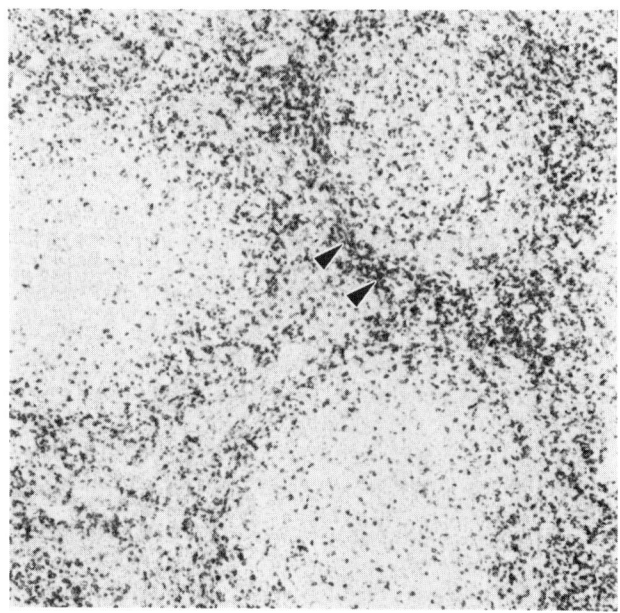

Figure 10. Methylmethyacrylate section of tonsil prepared as in *Protocol 2* stained with the antibody UCHL1 for the demonstration of T cells (arrows) using the streptavidin–biotin technique (*Protocol 4*).

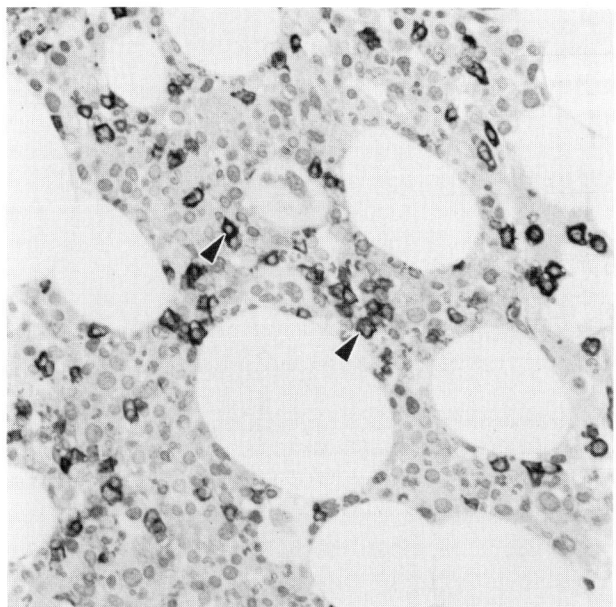

Figure 11. Methylmethacrylate section of bone marrow trephine prepared as in *Protocol 2* stained with the anti-Common Leucocyte Antigen (CD45) antibody using the streptavidin–biotin technique (*Protocol 4*).

6. Cytological preparation techniques

Cytological preparations possess several advantages over histological sections:

(a) Cytology specimens can be easily divided, allowing numerous smears/cytospins to be prepared. This permits the use of different fixatives, preparative methods, and pre-treatments.

(b) Fixation is rapid and easily controlled.

(c) Prompt fixation of specimens is not as critical as for paraffin sections and specimens may be allowed to stand overnight at 4 °C enabling the optimum fixation and preparative methods to be decided.

(d) Preparation of smears/cytospins is both quick and easy.

(e) Ample material is available, as smears/cytospins do not require large numbers of cells.

Cytological specimens however do have three disadvantages:

(a) lack of tissue architecture

(b) high background staining with some antigens on certain specimens

(c) poor condition of cells from certain specimens

Immunostaining can be achieved on the whole range of cytological specimens. These include pleural and ascitic fluids, urine, sputum, cerebrospinal fluid, washings from the gastrointestinal and respiratory tracts, gynaecological smears, and fine-needle aspirates.

6.1 Preparative methods

The preparative method will depend upon the type of specimen and the antigen to be demonstrated.

6.1.1 Smears/cytospins

(a) For liquid specimens centrifuge the specimen at 120*g* for 10 min.

(b) Drain off the supernatant, re-suspend the pellet in Tris-buffered saline 0.5 M pH 7.6 (TBS, see Appendix).

(c) Prepare smears manually on APES-coated slides. Alternatively prepare cytospins using a Shandon Cytospin:
   i. place 2–3 drops of re-suspended cells per cytospin chamber
   ii. centrifuge at 120*g* for 10 min

(d) Air-dry smears/cytospins. Some antigens may show decreased immunostaining after air-drying and in these cases omit this step.

(e) Fix smears/cytospins as desired. The most commonly used fixatives are ethanol and methanol (often combined), acetone, or formalin-based fixatives.

(f) Immunostain using the method of choice. The use of the APAAP method helps to overcome the problem of endogenous staining, which is sometimes evident on smears/cytospins stained by peroxidase methods.

The following points should be noted:

(a) Cytospin preparations have the advantage of producing an even spread of cells over a small area, hence small volumes of reagents are required for immunostaining.

(b) Specimens from high risk patients should be fixed before preparing cytospins.

(c) For the demonstration of lymphoid antigens, smears/cytospins should be air-dried and fixed in acetone for 20 min.

(d) Air-dried preparations can be stored by wrapping the slides in aluminium foil and storing at −20 °C.

(e) For the demonstration of intracellular antigens the use of detergents, such as Tween 20 (0.01% in TBS) may be required. Detergents permeabilize the plasma membrane allowing the antibody access into the cell. This treatment is not necessary when smears/cytospins have been fixed in alcohol or acetone, or have been air-dried.

6.1.2 Cell blocks

Small fragments of tissue are often removed during fine-needle aspiration cytology and it is useful to examine these pieces of tissue by preparing cell blocks.

Protocol 3. The preparation of sections of specimens embedded in agarose for immunolabelling

1. Fix tissue fragments or cells in 10% formalin.
2. Place in centrifuge tube and centrifuge at 120g for 10 min.
3. Decant the supernatant.
4. Pipette molten 4% agarose on to the tissue/cell pellet. Cover to a depth of 3 mm.
5. Centrifuge at 260g for 30 sec.
6. Allow agarose to set. Cover agarose 'button' with 10% formalin to assist in the setting.
7. Gently remove agarose 'button', and process to paraffin wax or resin.
8. Cut sections and immunostain as for paraffin/resin sections.

The recently developed Shandon Cytoblock (Shandon Southern Products, UK) technique has in our laboratory given good, consistent results, and is easier to use than the agarose method in *Protocol 3*.

6.1.3 Dabs/imprints

Dabs, or imprints, can be made from fresh tissue before fixation. Tissue is gently placed, cut surface down, on to APES-coated slides. The tissue is lifted away without smearing. This is repeated for the required number of slides. The dabs/imprints are appropriately fixed (generally with alcohol) prior to immunostaining.

The use of dabs/imprints allows rapid immunocytochemistry on gross histological specimens before paraffin sections have been prepared.

7. Immunolabelling techniques

There are several different techniques for immunolabelling the specimen:

- streptavidin–biotin technique (*Protocols 4* and *5*)
- direct technique (*Protocol 6*)
- indirect technique (*Protocol 7*)
- Protein A technique (*Protocol 8*)

- peroxidase–anti-peroxidase technique (*Protocol 9*)
- alkaline phosphatase–anti-alkaline phosphatase technique (*Protocol 10*)
- immunogold–silver staining technique (*Protocol 11*)

The selection of a particular technique will depend on the availability of the reagents, whether or not multiple immunolabelling (see Chapter 7) is to be performed and, to a certain degree, on the preference of the investigator.

Protocol 4. The streptavidin–biotin technique for paraffin sections

1. De-wax sections (including positive and negative controls) thoroughly in xylene.
2. Rinse in absolute alcohol.
3. Treat with 0.5% hydrogen peroxide in methanol (8 ml of 30% w/v hydrogen peroxide in 400 ml of methanol) for 30 min or with 3% hydrogen peroxide in deionized water for 5 min.
4. Wash in water.
5. Trypsinize sections if necessary (see Section 5.1 above) as follows:
   (a) Place sections in pre-warmed (37 °C) distilled water.
   (b) Prepare trypsin solution, 0.1% trypsin in 0.1% calcium chloride:
   - pre-warmed distilled water   400 ml
   - trypsin (Difco)   200 mg
   - 5% calcium chloride   8 ml
   - adjust to pH 7.8 with 1% sodium hydroxide.
   (c) Incubate sections for the required time.
6. Wash in cold water, to prevent further digestion.
7. Rinse sections in Tris-buffered saline pH 7.6 (TBS).
8. Dry excess buffer from around the sections and place in humidity chamber. Cover sections with a 1 in 5 dilution of normal serum from the animal species used to prepare the secondary antibody in step **11**. Incubate for 10 min.
9. Wipe off excess normal serum. Cover sections with optimally diluted primary antibody. Dilute antibodies in 0.01% Tween 20 in TBS (TTBS). Incubate polyclonal antibodies for 1 h (minimum), incubate monoclonal antibodies overnight.
10. Wash well in TBS.
11. Wipe off excess TBS. Incubate sections in appropriate secondary antibody for 30 min:

(a) For polyclonal antibodies raised in rabbit, use 1:200 biotinylated goat anti-rabbit immunoglobulin (Vector BA-1000 USA, UK).

(b) For monoclonal antibodies raised in mouse use 1:200 biotinylated rabbit anti-mouse immunoglobulin (Vector BA-2000 USA, UK).

Dilute secondary antibodies with TTBS.

12. Wash sections well in TBS.

13. Prepare streptavidin–biotin/horseradish peroxidase complex (StrABC/ HRP) (Vectastain Elite PK-6102 USA, UK):

- 20 µl streptavidin
- 20 µl biotinylated HRP
- 1 ml TTBS

Mix well and allow to stand for 30 min.

14. Wipe off excess TBS and incubate sections in StrABC/HRP for 30 min.

15. Wash well in TBS.

16. Develop peroxidase in DAB solution as follows:

(a) To 400 ml of Tris buffer add 3 ml of stock DAB solution (see the Appendix) and 12 drops of hydrogen peroxide (30% w/v).

(b) Incubate with sections, checking with a microscope after 10 min. Development times, may be extended if necessary.

17. Wash in water.

18. Incubate sections in copper sulphate solution for 5 min (see the Appendix).

19. Wash in water and counterstain as desired.

20. Dehydrate, clear, and mount in synthetic mounting medium.

Protocol 5. The streptavidin–biotin technique for frozen sections

1. Rehydrate frozen sections (including positive and negative controls) in TBS.

2. Wipe off excess TBS, incubate sections for 1 h with primary antibody optimally diluted with TBS.

3. Wash well in TBS.

4. Wipe off excess TBS, incubate sections in appropriate secondary antibody for 30 min:

(a) For polyclonal antibodies use 1/200 biotinylated goat anti-rabbit immunoglobulin.

Protocol 5. *Continued*

    (b) For monoclonal antibodies raised in mouse use 1/200 biotinylated rabbit anti-mouse immunoglobulin.

5. Wash sections in TBS.

6. Prepare streptavidin–biotin/horseradish peroxidase complex (StrABC/HRP) as follows:

    ● 20 μl of streptavidin

    ● 20 μl of biotinylated HRP

    ● 1 ml of TBS

    Mix well and allow to stand for 30 min.

7. Wipe off excess TBS, incubate sections in StrABC/HRP for 30 min (minimum).

8. Wash well in TBS.

9. Develop as for the StrABC (*Protocol 4*, step **16**).

10. Wash in water.

11. Incubate in copper sulphate solution for 5 min (see the Appendix).

12. Wash in water. Counterstain as desired.

13. Dehydrate, clear, and mount in synthetic mounting medium.

Notes

(a) Blocking endogenous peroxidase activity is not usually necessary and indeed the effect of methanol/H_2O_2 may destroy or alter antigenic sites in frozen sections. If necessary use a mild technique (Chapter 10, Section 3.2.1.ix).

(b) Sections are fixed in acetone (*Protocol 1*) before rehydration and therefore treatment with proteolytic enzymes is not necessary.

(c) Antibodies are diluted in TBS. The use of Tween 20 should be avoided as it may increase the detrimental effects of chromatolysis and loss of nuclear membranes.

Protocol 6. Direct technique

1. Follow the streptavidin–biotin technique until step **8** for paraffin sections (*Protocol 4*), or step **1** for frozen sections (*Protocol 5*).

2. Wipe off excess normal serum or TBS, incubate sections in appropriate peroxidase-conjugated primary antibody diluted with PBS. Incubate sections for 1 h.

3. Wash well in TBS.

4. Develop as for the StrABC techniques (*Protocol 4*, step **16**).

5. Enhance in copper sulphate solution for 5 min (see the Appendix).

6. Counterstain as desired.

7. Dehydrate, clear, and mount.

Protocol 7. The indirect technique

1. Follow the streptavidin–biotin technique until step **10** for paraffin sections (*Protocol 4*) or step **3** for frozen sections (*Protocol 5*).

2. Wipe off excess TBS and incubate sections in appropriate peroxidase-conjugated secondary antibody for 30 min:
   (a) For polyclonal antibodies raised in rabbit use 1/50 swine anti-rabbit immunoglobulin/HRP (Dako P162 Denmark, UK, USA).
   (b) For monoclonal antibodies raised in mouse 1/50 rabbit anti-mouse immunoglobulin/HRP (Dako P161 Denmark, UK, USA).

3. Wash well in TBS.

4. Develop as for the StrABC technique (*Protocol 4*, step **16**).

5. Enhance in copper sulphate solution for 5 min (see the Appendix).

6. Counterstain as desired.

7. Dehydrate, clear, and mount.

Protocol 8. The Protein A technique

1. Follow the streptavidin–biotin technique until step **7** for paraffin sections (*Protocol 4*) or step **3** for frozen sections (*Protocol 5*).

2. Block non-specific binding sites with a 1 in 5 dilution of globulin-free[a] bovine serum albumin (BSA). Block for 5 min.

3. Drain and incubate sections with optimally diluted primary antibody, diluted in 1 in 20 globulin-free BSA in TBS pH 7.6.

4. Wash well in TBS.

5. Incubate with optimally diluted Protein A–peroxidase conjugate for 1 h.

6. Wash well in TBS.

7. Develop as for the StrABC technique (*Protocol 4*, step **16**).

8. Enhance in copper sulphate solution for 5 minutes (see the Appendix).

Protocol 8. *Continued*

9. Counterstain as desired.
10. Dehydrate, clear, and mount.

ᵃ Globulin-containing blocking agents should not be used as non-specific Protein A–globulin reactions may occur

---

Protocol 9. The peroxidase–anti-peroxidase (PAP) technique

1. Follow the streptavidin–biotin technique until step **10** for paraffin sections (*Protocol 4*) or step **3** for frozen sections (*Protocol 5*).
2. Wipe off excess TBS and incubate sections in appropriate secondary antibody for 30 min:
   (a) For polyclonal antibodies raised in rabbit use 1/25 swine anti-rabbit immunoglobulins (Dako Z196 Denmark, UK, USA).
   (b) For monoclonal antibodies raised in mouse use 1/50 rabbit anti-mouse immunoglobulins (Dako Z259 Denmark, UK, USA).
3. Wash well in TBS.
4. Wipe of excess TBS and incubate with appropriate PAP reagent for 30 minutes:
   (a) For polyclonal antibodies raised in rabbit use 1/50 PAP (Dako Z113 Denmark).
   (b) For monoclonal antibodies raised in mouse use 1/50 mouse PAP (Dako P850 Denmark, UK, USA).
5. Wash well in TBS.
6. Develop as for the StrABC technique (*Protocol 4*, step **16**).
7. Enhance in copper sulphate solution for 5 min (see the Appendix).
8. Counterstain as desired.
9. Dehydrate, clear, and mount.

Protocol 10. The alkaline phosphatase–anti-alkaline phosphatase (APAAP) technique using a mouse monoclonal antibody

1. Follow the streptavidin–biotin technique until step **10** for paraffin sections (*Protocol 4*) or step **3** for frozen sections (*Protocol 5*).
2. Wipe off excess TBS, incubate section for 1 h into 1/50 rabbit anti-mouse immunoglobulin (Dako Z259 Denmark, UK, USA) diluted with TBS.

3. Wash well in TBS.

4. Wipe off excess TBS, incubate section for 1 h in 1/10 mouse APAAP (Dako D651 Denmark, UK, USA) diluted with TBS.

5. Wash well in TBS.

6. Prepare the developing solution as follows:

   (a) Dissolve 10 mg naphthol AS/MX phosphate in 1 ml of dimethyl formamide.

   (b) Add 49 ml of 0.1 M Tris buffer pH 8.2.

   (c) Immediately before use add 50 mg Fast Red TR or 50 mg Fast Blue BBN, and 25 mg of levamisole.[a]

7. Incubate sections with developing solution in a coplin jar for 30 min at 37 °C.[b]

8. Wash well in water.

9. Counterstain if desired.

10. Mount from water in pre-warmed glycerine jelly. Ring sections with glyceel (BDH, UK) when dry.

[a] The addition of levamisole is to inhibit endogenous alkaline phosphatase.
[b] An alternative, permanent red end product may be achieved using the following substrate solution at step 7:
(a) Dissolve 10 mg of naphthol AS-TR phosphate in 200 µl of dimethylformamide.
(b) Add 40 ml of 0.2 M Tris buffer pH 9.0.
(c) Immediately before use mix:
  ● 250 µl of 4% basic fuchsin in 2M HCl
  ● 250 µl of 4% sodium nitrite in distilled water
  Allow to stand for 1 min.
(d) Add the hexazotized basic fuchsin to the substrate solution, add 25 mg of levamisole. Filter before use.
(e) Develop at room temperature for 10 min.
(f) Wash well in water.
(g) Counterstain if desired.
(h) Dehydrate quickly, clear, and mount in synthetic mounting medium.

Protocol 11. The immunogold–silver staining technique

1. Follow the streptavidin–biotin technique for paraffin sections until step 7 (*Protocol 4*).

2. Immerse sections in Lugol's iodine for 5 min.

3. Wash in water and immerse sections in 2.5% sodium thiosulphate solution.

Protocol 11. *Continued*

4. Rinse in TBS.

5. Block non-specific background staining with normal goat serum (NGS), for 5 min.

6. Wipe off excess NGS and incubate sections for 1 h in primary antiserum diluted to optimal concentration with 1/20 NGS in TBS.

7. Wash sections in TBS for 10 min.

8. Repeat NGS block for 5 min.

9. Apply appropriate immunogold reagent (1 nm or 5 nm gold spheres coated with appropriate anti-Ig molecules) diluted 1/20 in TBS pH 8.2, containing 1% BSA for 2 h.

10. Wash in TBS for 10 min, and in distilled water for 10 min.

11. Develop sections in silver solution for up to 30 min in a darkroom using Ilford Safelight S902 or F904. Examine the sections periodically to determine optimum development time.[a] The silver solution has the composition

    - gum acacia (500 g/litre)   7.5 ml
    - distilled water   52.5 ml
    - citrate buffer pH 3.5   10 ml
    - hydroquinone[b] (0.8 g/15 ml)   15 ml
    - silver lactate[b,c] (0.11 g/15 ml)   15 ml

    All solutions are made in distilled water and mixed in the above order.

12. Rinse sections in 2.5% sodium thiosulphate. Wash thoroughly in distilled water.

13. Counterstain as desired.[d]

14. Dehydrate, clear, and mount in DPX.[e]

For frozen sections, rehydrate in TBS and proceed from step **5**. After step **10**, fix with 2% glutaraldehyde in TBS for 15 min and rinse for 10 min in distilled water before proceeding to step **11**.

---

[a] The silver intensification may be performed using one of the commercial silver enhancer kits. These are light insensitive and enhancement may be performed using normal laboratory lighting conditions. During silver enhancement, the process may be stopped at any time by washing in distilled water. The sections may be safely examined using light microscopy. If silver enhancement is not optimal the sections may be re-immersed in the enhancing solution.

[b] The hydroquinone and silver lactate solutions should be prepared immediately before use.
[c] All solutions containing silver lactate should be protected from light.
[d] Sites of positive immunostaining appear black.
[e] The end reaction product may fade with some synthetic mounting media but not in DPX.

8. Alternative substrates for immunoperoxidase techniques

If the need arises, such as for multiple immunolabelling the peroxidase may be developed with different substrates to produce different coloured reaction products.

Protocol 12. Development of peroxidase with 3-amino-9-ethyl-carbazole[a]

1. Follow the immunoperoxidase technique of choice until the development stage (e.g. *Protocol 4*, step **16**).

2. Incubate sections for 2–5 min at room temperature in the substrate solution prepared just before use as follows:

   • Dissolve 60 mg of 3-amino-9-ethylcarbazole in 15 ml of dimethyl-formamide.

   • Add 285 ml of 0.5 M acetate buffer pH 5.2 (see the Appendix) (it is not necessary to re-adjust the pH).

   • Immediately before use, add 1 ml of 3% hydrogen peroxide solution.

3. Wash well in water.

4. Mount from water in pre-warmed glycerine jelly. Ring sections with glyceel when dry.

[a] 3-amino-9-ethylcarbazole gives a red end product.

Protocol 13. Development of peroxidase with 4-chloro-1-naphthol[a]

1. Follow the immunoperoxidase technique of choice until the development stage (e.g. *Protocol 4*, step **16**).

2. Incubate sections for 30–90 min at room temperature in the substrate solution prepared as follows:

   • Dissolve 120 mg of 4-chloro-1-naphthol in 1.5 ml of absolute ethanol, place at −20 °C for 30 min.

   • To 300 ml of TBS (pH 7.6) add 250 μl of hydrogen peroxide.

   • Mix the two solutions, filter, and use immediately.

3. Wash well in water.

Protocol 13. *Continued*

4. Mount from water in pre-warmed glycerine jelly. Ring sections with glyceel when dry.

ª 4-chloro-1-naphthol gives a blue end product.

Appendix

A.1 Specialized fixatives

(a) B5 (stock solution):

mercuric chloride	60 g
sodium acetate	12g
distilled water	1 litre

Working solution:	
B5 stock solution	100 ml
formaldehyde (40% w/v)	10 ml

(b) Periodate–lysine–paraformaldehyde dichromate (PLP/D):

3% (w/v) paraformaldehyde	100 ml
0.1 M disodium hydrogen orthophosphate	200 ml
lysine	1.8 g
sodium periodate	0.3 g
adjust to pH 7.4	

(c) Periodate–lysine–paraformaldehyde dichromate (PLP/D):

3% (w/v) paraformaldehyde	100 ml
0.1 M disodium hydrogen orthophosphate	200 ml
lysine	1.8 g
sodium periodate	0.3g
potassium dichromate	15 g
adjust to pH 7.4	

A.2 Stock solutions

(a) Tris buffered saline (0.5 M) pH 7.6 (TBS):

  i. Dissolve 60.5 g of Tris (hydroxyl) methylamine in about 750 ml of distilled water.

  ii. Adjust to pH 7.6 with 50% hydrochloric acid.

  iii. Add 85 g of sodium chloride, make final volume to 10 litres with distilled water.

(b) Diaminobenzidene (DAB) solution:

  i. Dissolve 7.5 g of 3,3-diaminobenzidene tetrahydrochloride in 300 ml of Tris buffer pH 7.6.

ii. Aliquot into 1 ml amounts and store in deep freeze. DAB is a suspected carcinogen. Appropriate care must be taken when handling this reagent.

(c) Copper sulphate solution:

copper sulphate	4 g
sodium chloride	7.2 g
distilled water	1 litre

(d) 0.05 M acetate buffer pH 5.2:

0.1 M acetic acid	63 ml
0.1 M sodium acetate	237 ml

A.3 Aminopropyltriethoxysilane (APES) coated slides

(a) Place clean slides in a staining rack, wash in acetone.

(b) Allow to drain, immerse in 2% APES in acetone.

(c) Allow to drain (but not dry), wash in running water.

(d) Allow to air-dry.

References

1. Coons, A. H., Creech, H. J., Jones, R. N., and Berliner, E. (1942). *J. Immunol.*, **45**, 159.
2. Nakane, P. K., and Pierce, G. B. (1966). *J. Histochem. Cytochem.*, **14**, 929.
3. Avrameas, S., and Uriel, J. (1966). *C.R. Acad. Sci. Paris*, **262** 2453.
4. Foresgren, A., and Sjöquist, J. (1966). *J. Immunol.*, **97**, 822.
5. Roth, J. (1983). In *Techniques in Immunocytochemistry*, vol. 2, (ed. G. R. Bullock and P. Petrusz), pp. 217–84. Academic Press, London.
6. Faulk, W., and Taylor, G. (1971). *Immunochemistry*, **8**, 1081.
7. Geoghegan, W. D., and Ackerman, G. A. (1977). *J. Histochem. Cytochem.*, **25**, 1187.
8. Gu, J., De Mey, J., Moremans, M., and Polak, J. M. (1981). *Reg. Peptides*, **1**, 365.
9. Danscher, G. (1981). *Histochemistry*, **71**, 81.
10. Holgate, C. S., Jackson, P., Cowen, P. N. and Bird, C. C. (1983). *J. Histochem. Cytochem.*, **31**, 933.
11. De Mey, J. (1983). In *Immunocytochemistry, Practical Applications in Pathology and Biology*, (ed. J. M. Polak and S. Van Noorden), pp. 82–112. Wright, Bristol.
12. De Waele, M., De Mey, J., Renmans, W., Labeur, C., Jochmans, K., and Van Camp, B. (1986). *J. Histochem. Cytochem.*, **34**, 1257.
13. McLean, I. W. and Nakane, P. K. (1974). *J. Histochem. Cytochem.*, **22**, 1077.
14. Holgate, C. S., Jackson, P., Pollard, K., Lunny, D., and Bird, C. C. (1986). *J. Pathol.*, **149**, 293.
15. Huang, S., Minassian, H., and More, J. D., (1976). *Lab. Invest.*, **35**, 383.
16. Hand, M. N. (1989). *Proc. R. Microsc. Soc.*, **24**, A54.

4

Immunolabelling techniques for electron microscopy

PAUL MONAGHAN[a], DAVID ROBERTSON[a], and
JULIAN E. BEESLEY[b]

A. The post-embedding techniques
Paul Monaghan and David Robertson

1. Introduction

Electron microscope (EM) immunolabelling techniques can be broadly divided into two groups. Those where the immunolabelling steps take place on (usually) unfixed and non-embedded samples are referred to as pre-embedding and are covered elsewhere in this chapter (Sections 6–9, parts B and C). Those methods where the labelling is undertaken on (usually) fixed, embedded, and sectioned material are known as post-embedding.

The choice of whether to apply pre- or post-embedding methods to the detection of an antigen in any particular location will depend to a large extent upon the distribution and lability of the antigen and the characteristics of the primary antibody. Unless the antigen is readily accessible in the unprocessed sample, a post-embedding method will almost certainly be required.

Conventional processing protocols provide excellent morphological preservation, but fixation, dehydration, infiltration, and heat polymerization are all capable of modifying antigens such that they are unrecognized by their antibody. Unfortunately almost every antigen–antibody combination responds to these various insults to a greater or lesser extent and it is not possible *de novo* to predict the ease with which an antigen can be detected in any situation. Whilst some antigens are demonstrable in glutaraldehyde and even osmium-fixed and epoxy resin-embedded samples (1, 2) these are in the minority and a number of methods have been devised to reduce processing damage to antigens.

Four main methods have been devised for post-embedding immunolabelling

---

[a] Authors of part A, Sections 1–5.
[b] Author of parts B and C, Sections 6–9.

and will be covered in this chapter. Each has its advantages and disadvantages which will be discussed in the relevant sections.

The methods covered are:

- embedding at ambient temperature in LR White
- embedding in low temperature resin by the progressive lowering of temperature method (PLT)
- freeze-substitution
- thawed cryosections

Taking the three major areas of processing with the potential to damage antigens as fixation, dehydration, and resin embedding, it is not yet possible to avoid all three processes and still be able to apply immunolocalization procedures. As summarized in *Table 1* it is possible to avoid or minimize the effects of some of the steps and therefore, with the correct approach, most antigens can be detected.

The methods that will be described have been in use in our laboratory for many years, and are a distillation of much published literature with a number of our own modifications. Whilst not exhaustive, they should suffice for the majority of studies and will make a good starting point for the more adventurous. They have been successfully applied to a large number of antigens and a few examples are illustrated.

How is the most appropriate method chosen? If at all possible, the characteristics of the antibody should be determined before starting any ultrastructural work. This is best performed at the light microscope level and worked through to the final successful labelling of EM sections.

Table 1. Summary of the major steps in post-embedding immunocytochemistry with the potential to damage antigens

| | **Potentially antigenically damaging steps** | | |
Method	Fixation	Dehydration/ substitution	Resin embedding
LR White	+	+	+
PLT	+	+	+
Freeze-substitution	−	+	+
Thawed cryosections	+	−	−

2. Light microscope characterization of reagents

Protocol 1. Preliminary light microscope immunolabelling

1. Cut a series of 7 μm frozen sections of the tissue under investigation and place on glass microscope slides.

2. Fix for 30 min, 1 h, and 2 h with phosphate-buffered saline (PBS) solutions of
   - 2% glutaraldehyde
   - 2% paraformaldehyde plus 0.05% glutaraldehyde (2.05 fix)
   - 4% paraformaldehyde
3. Immunolabel using the antibody in question with either peroxidase or alkaline phosphatase conjugated second antibody (Chapter 3). Alternatively the second antibody can be a colloidal gold conjugate visualized by silver enhancement.

If, when compared with unfixed control sections, the fixed sections indicate that a significant proportion of the antigen is still present after 1 h in 2.05 fixative, then all four methods (LR White, PLT, thawed cryosections, and freeze-substitution) may prove successful. If, however, the antigen is rapidly lost with even short fixation times then freeze-substitution may present the only method of post-embedding labelling available.

There is some evidence from light microscope immunocytochemistry that different dehydration solvents may have different effects upon antigens (3). For low-abundance or particularly sensitive antigens, the LM experiment determining the relative effects of fixatives on immunolabelling should be repeated with:

- acetone
- methanol
- ethanol

The results of these experiments will demonstrate to a considerable degree the characteristics of the antigen/antibody combination, and allow an informed choice of processing approach which will improve the chances of a successful localization.

3. Resin embedding techniques

Resin embedding techniques for immunocytochemistry fall into two main areas. These are embedding at ambient temperature and low temperature embedding.

3.1 Ambient temperature

For embedding at ambient temperature, LR White resin is generally used in preference to epoxy resins. LR White also has the advantage of being able to infiltrate partially hydrated specimens, so that samples can be transferred to resin from 70% alcohol dehydrating agents (4). This may have implications for the retention of antigenicity of some antigens.

Protocol 2. Embedding in LR White

1. Fixation (see Section 3.2.1).
2. Wash overnight in buffer of choice.
3. Transfer to 70% ethanol for 60 min.
4. Perform two further changes of 70% ethanol for 60 min.
5. Transfer to LR White and agitate for 60 min.
6. Transfer to fresh LR White overnight.
7. Embed in fresh LR White in gelatin capsules.
8. Polymerize at 50 °C for 24 h under anaerobic conditions obtained by filling the capsule with resin and closing with the lid.

Whilst the use of LR White at ambient temperature is rapid and convenient there is no doubt that embedding at low temperatures provides improved morphological preservation over ambient temperature techniques, and here the Lowicryl resins excel. The following protocols will therefore deal with low temperature embedding methods.

Two lower temperature methods will be described for the preparation of resin-embedded samples. These are low temperature embedding into Lowicryl resin by PLT and freeze-substitution. PLT embedding has been devised to reduce the denaturation and extraction of proteins in fixed samples by undertaking the dehydration and embedding stages at progressively lowered temperatures. The dehydration is followed by infiltration, embedding, and UV polymerization of the resin also at low temperature. Clearly it is only appropriate for antigens which withstand at least a brief fixation. Freeze-substitution on the other hand can be applied with or without fixatives in the substitution medium, but for fixation-sensitive antigens exposure to cross-linking fixatives can be avoided.

3.2 Progressive lowering of temperature

The attributes of this method can be briefly summarized as follows:

(a) advantages:
- good morphology
- permanent blocks
- relatively easy and cheap

(b) disadvantages:
- aldehyde fixation needed
- dehydration and embedding may affect antigens

The method consists of:

- brief aldehyde fixation
- dehydration at low temperature
- infiltration in Lowicryl resin
- UV polymerization at low temperature

3.2.1 Fixation

Different antigens are affected by cross-linking fixatives in different ways. Some withstand long periods in glutaraldehyde-based fixatives and are still readily demonstrable, whilst other antigens are rendered unrecognizable by brief exposure to low concentrations of paraformaldehyde. In between these extremes are many antigens where some epitopes appear more sensitive to fixation than others. For instance, in a series of monoclonal antibodies raised against a single molecule, some may react with fixed antigen whilst others will not. The message here is that where available it may be worth investigating more than one antibody/antiserum if fixation sensitivity seems to be a problem.

With the above comments on the differing reactions of antigens and antibodies following exposure to cross-linking fixatives, hard and fast rules cannot be made. The following comments on fixation are derived from published data and experience in our laboratory with a wide range of applications.

Glutaraldehyde at high concentrations (> 1%) should, in general, be avoided. Paraformaldehyde, freshly prepared, is less damaging to many antigens, but for tissue culture cells in particular, 4% paraformaldehyde can give poor preservation. A routine fixative for immunocytochemistry consisting of 2% paraformaldehyde plus 0.05% glutaraldehyde (2.05) in PBS gives a good compromise between retention of morphology and antigenicity.

3.2.2 Duration of fixation

Prolonged fixation should be avoided as many antigens show a progressive reduction in immunoreactivity with increasing fixation times. Although fixation for less than 60 min may be inadequate for anything but the smallest sample, fixation in 2.05 for 60 min will be optimal for many situations.

If not for immediate use, the samples should be stored in PBS and not in the fixative. Long term storage in buffer cannot be recommended as a routine method, however, as fixation with paraformaldehyde is to some extent reversible. The time course of this process is not clear, and therefore embedding should follow fixation with minimal delay. In our laboratory, samples have been fixed for 1 h in 2.05, stored in PBS for up to one month and subsequently processed and successfully immunolabelled.

3.2.3 Dehydration and embedding

The protocol to be followed will vary with the choice of resin used for embedding. The Lowicryl resins were formulated to provide embedding media with low viscosity for the preservation of the molecular and antigenic structure of the cell (5). They consist of four variants which differ in their hydrophobic or hydrophilic nature, and the temperature at which they can be used (*Table 2*). All Lowicryl resins have low viscosity even at low temperatures and are polymerized by long wavelength UV light.

It was initially thought that the hydrophilic nature of K4M and K11M would assist the immunocytochemical reactions at the resin surface, and many publications have demonstrated the suitability of K4M in particular for immunocytochemistry. Recent work, however, indicates that the sensitivity of the different resins is not markedly different and it is the surface relief of the sectioned resin that affects the quantity of immunolabelling on a section (6). The epoxy resins produce sections which are relatively smooth whereas Lowicryl resin sections show greater surface irregularities. This surface irregularity is postulated to expose antigenic sites at the surface of the section. In some situations, the quality of preservation obtained with HM20 is superior to that seen with K4M (7) and in the following protocols HM20 will be used. The principles remain the same for all the Lowicryl resins and, apart from the different minimum temperatures, can be treated in an identical manner.

Table 2. Characteristics of Lowicryl resins

Resin	Nature	Useable down to
K4M	Hydrophilic	$-35\,°C$
HM20	Hydrophobic	$-50\,°C$
K11M	Hydrophilic	$-60\,°C$
HM23	Hydrophobic	$-80\,°C$

3.2.4 Provision of different temperatures

The dehydration protocol for HM20 requires the provision of temperature points of:

- $0\,°C$
- $-15\,°C$
- $-30\,°C$
- $-50\,°C$

These may be provided in a number of ways. Standard laboratory refrigerators and freezers may well be available for these temperature points, but some ingenuity may be required if not available. For example, sodium chloride/ice mixtures will approximate to the $-15\,°C$ point. A simple aluminium holder can be produced for the transfers at each temperature and acts as a heat sink to minimize temperature drift (*Figure 1*). Alternatively, all stages can be undertaken in a Balzers (Germany and UK) low temperature embedding unit (LTE 020) or the Leica CSAuto (UK). This latter device, although ideal for freeze-substitution is also suitable for PLT as it is temperature controlled and equipped with a UV light source for polymerization.

Figure 1. Aluminium block which acts as a heat-sink during transfers in the dehydration stages of PLT embedding. A small quantity of methanol is added to the holes in the holder to assist thermal conductivity with the specimen pots.

Protocol 3. Fixation and dehydration for PLT embedding

1. Fix samples in 2.05 for 60 min.

2. Wash in PBS for 60 min minimum.

3. Transfer to 30% ethanol at 0 °C for 30 min.

4. Transfer to 55% ethanol at −15 °C for 60 min.

5. Transfer to 70% ethanol at −30 °C for 60 min.

6. Transfer to 100% ethanol at −50 °C overnight.

7. Transfer to 100% ethanol at −50 °C for 60 min.

3.2.5 Resin infiltration

Prepare the Lowicryl resin in the ratios recommended by the manufacturer. Mix by gently bubbling nitrogen gas through the resin. This should be performed in a fume cupboard. It reduces the oxygen content of the resin and provides a layer of nitrogen above the resin thereby reducing oxygen contact with the resin. This is important as resin polymerization is inhibited by the presence of oxygen. As soon as polymerization is complete, removal of the capsules and subsequent exposure to the oxygen in the atmosphere will prevent further polymerization during storage of the blocks.

Protocol 4. Infiltration and embedding

Following dehydration (*Protocol 3*), samples are infiltrated in HM20 by the steps outlined. The infiltration steps have been increased to ensure complete infiltration of the samples.

1. Infiltrate with a mixture of 25% resin: 75% ethanol for 60 min.
2. Change to a mixture of 50% resin: 50% ethanol for 60 min.
3. Transfer to a mixture of 75% resin: 25% ethanol for 60 min.
4. Infiltrate in 100% resin for 60 min.
5. Repeat 100% resin overnight.
6. Finally infiltrate in 100% resin for 60 min.
7. Embed in gelatin or plastic (UV transparent) capsules.
8. Polymerize in an anaerobic environment with UV light for 48 h at −50 °C and finally 24 h at 0 °C.

Depending on the UV intensity and the distance from the resin this timing may need to be varied to obtain well polymerized blocks. The UV polymerization of Lowicryl is an exothermic reaction, and without an adequate heat sink there may be a risk of an uncontrolled temperature rise during polymerization. For polymerization in the CSAuto (Leica, UK), the flat embedding moulds contain small volumes of Lowicryl, and for plastic embedding capsules, an aluminium block can be produced to fit in the instrument which acts as a large heat sink. Both approaches minimize the risk of temperature rise, and a maximum of 2 °C temperature rise during polymerization was recorded with both HM20 and K11M (Robertson, unpublished data).

Examples of PLT-embedded human kidney and breast tissue are illustrated in *Figures 2* and *3*.

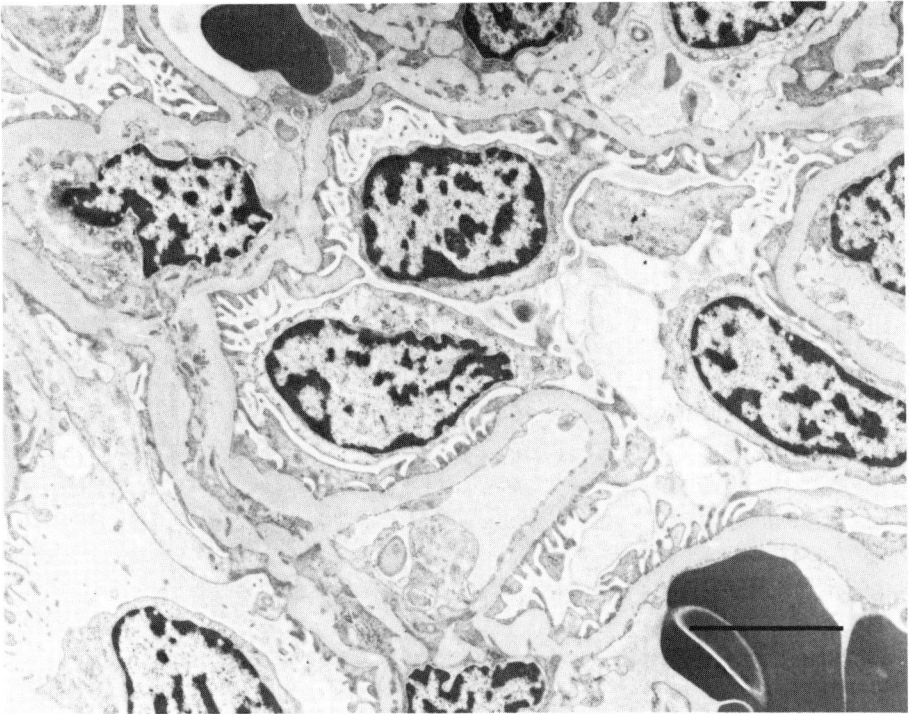

Figure 2. Human kidney fixed for 1 h in 2% paraformaldehyde plus 0.05% glutaraldehyde fixative and embedded in Lowicryl HM20 by progressive lowering of temperature. Sections were contrasted with uranyl acetate and lead citrate. Scale bar 4 μm.

3.3 Freeze-substitution

Although an alternative resin embedding method, it differs from PLT embedding in several key areas. These centre upon the use of rapid freezing as an alternative to aldehyde fixation. Primary fixation with aldehydes is a relatively slow process, and rapid cellular events may not be observed in samples thus fixed. Rapid freezing stabilizes the cellular components within microseconds and preserves the cell in as near the *in vivo* state as possible. The features of the method are summarized as:

(a) advantages:

- permanent blocks
- good morphology in the well frozen region
- no aldehyde fixatives

(b) disadvantages:

- only small regions well frozen
- substitution and embedding may affect antigens

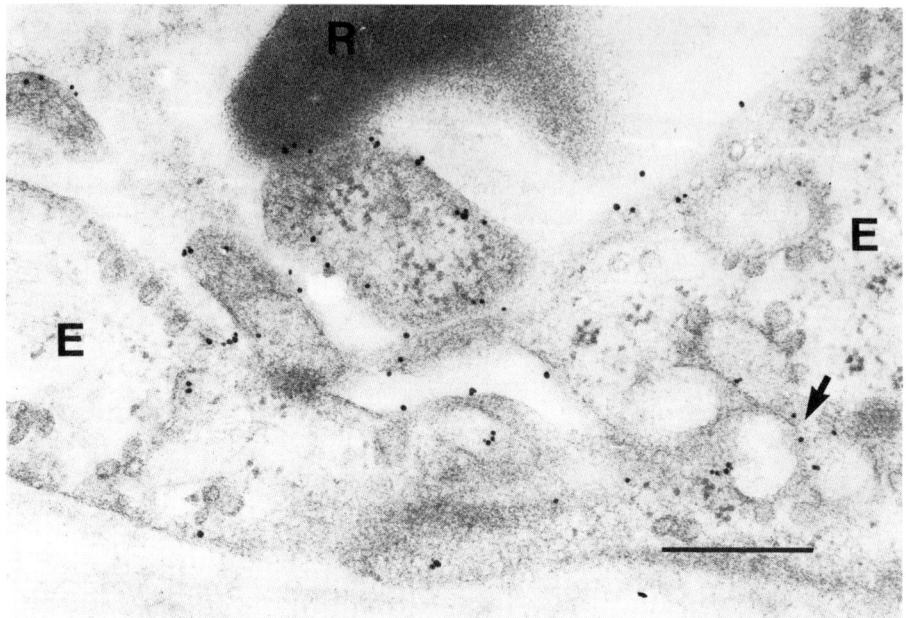

Figure 3. Human breast capillary processed as in *Figure 2*, and immunolaballed with anti-CD34 antibody. The 5 nm colloidal gold conjugate was silver enhanced for 6 min. Labelling is present on the membrane of two endothelial cells (E) and associated with cytoplasmic vesicles (arrow). The capillary contains a red blood cell (R). Scale bar 0.3 μm.

Freeze-substitution was applied initially to morphological studies and the investigation of dynamic cellular processes which are not preserved by chemical fixation. In such studies, the ice in the frozen sample is substituted by acetone or methanol containing glutaraldehyde, osmium tetroxide, uranyl acetate, or some combination of these. When the ice has been replaced, the sample is warmed to room temperature and embedded by routine methods (8). More recently, by omitting additives to the substitution medium, the method has been applied to immunocytochemical problems following atmospheric pressure (7) and high pressure freezing (9).

3.3.1 Freezing
Freezing rates of the order of 10^6 °C/sec are necessary to prevent the formation of ice crystals in the tissue. Such high freezing rates can be obtained at ambient pressure by several methods; the two commonly used ones being plunging into liquid cryogen cooled by liquid nitrogen (LN_2) and impact on to an LN_2-cooled copper block. Commercial apparatus is available for both methods (Leica, UK; RMC, UK, USA) although it is possible to design and construct simple devices for this purpose (10). The protocols that follow have

been established using the Leica MM80 (UK) metal mirror freezing apparatus, but the principles remain the same for all freezing methods.

i. Solid tissues

Samples to be frozen should be obtained as rapidly as possible to avoid post-mortem changes. Care must therefore be taken at this stage that any trimming of the sample results in minimal mechanical damage to the sample and that the sample is kept in a moist chamber to prevent air-drying. Freezing should be undertaken within seconds. Large samples are difficult to embed and polymerize at low temperatures, so small slices should be prepared. Despite the high freezing rates that can be achieved by both plunging and impact freezing, only the outer 10–15 µm can be well frozen. Further away from the cryogenic surface, the freezing rate is limited by the conductivity of the sample and ice crystal artefacts are apparent.

After freezing, samples must be kept away from exposure to the atmosphere to prevent water condensation on the surface and care must be taken to handle the frozen samples only with LN_2-cooled instruments.

ii. Samples in suspension

Samples in suspension such as cells from tissue culture or microbiological samples require some support to keep the sample together during the substitution and embedding stages. A 4% solution of gelatin in suitable medium or PBS can be used.

Protocol 5. Freezing of cell samples in suspension.

1. Centrifuge sample at 500*g* for 5 min (this speed should be varied to suit other types of specimen).
2. Warm 4% solution of gelatin in PBS or medium until liquid.
3. Re-suspend sample in 4% gelatin, keeping it liquid.
4. Centrifuge at 500*g* for 5 min to obtain a concentrated sample.
5. Mount a nylon washer on the freezing device sample holder.
6. Add a droplet of the concentrated sample into the washer so that the meniscus just protrudes over the washer and freeze as for solid tissues.

A concentrated sample is necessary because the well frozen region will only extend 10–15 µm into the sample and it is advantageous to maximize the sample present within this region.

The substitution and embedding protocols are the same as for solid tissues.

It is possible to remove the frozen sample from the washer after freezing but there is a risk of damaging the well frozen region. It is easier to process the washer complete with its frozen sample through to embedding and polymerization. The presence of the washer aids handling and ultimately the orientation of the block for sectioning. Once the sample is orientated correctly the washer can be cut away.

3.3.2 High pressure freezing

The major limitation of freeze-substitution is the depth of freezing obtained with any of the freezing methods described. Increasing the pressure surrounding a sample immediately prior to the freezing process can inhibit the formation of ice crystals and allow greater depths of well preserved material to be prepared. The optimum conditions are achieved when the pressure is increased to 2100 bar prior to freezing with LN_2. These conditions are not readily produced and only one commercial apparatus is available, from Balzers (Germany, UK).

Under optimal conditions, high pressure freezing is able to freeze up to 400 μm of tissue, giving a considerable enhancement over atmospheric pressure systems (11).

Following high pressure freezing, the samples are treated as for other freezing methods.

3.3.3 Substitution

For immunocytochemical studies, the substitution can be prepared with fixatives in the solvent, but the effects of fixatives at these low temperatures are difficult to predict. It is simplest to undertake the substitution between −80 and −90 °C with pure solvents not containing fixatives. Both acetone and methanol may be used. Methanol substitutes rapidly whereas acetone is much slower, and care must be taken to use acetone dried over a molecular sieve as it is unable to substitute fully with more than 1% water content. For methanol a substitution for 30 h is almost certainly over extended for many samples but currently conveniently fits the protocol into the working day. For acetone, substitution times recommended by different authors vary from 2–8 days, with the majority opting for shorter rather than longer substitution protocols. With either solvent there is the potential to optimize the protocol to the specific sample under investigation. Substitution takes place in any convenient chamber that can be maintained at the required temperature but once again commercial equipment is available (Balzers Germany, UK and Leica UK). After substitution is complete, the sample is warmed to the temperature for infiltration and polymerization, which in the case of Lowicryl HM20 is −50 °C. The infiltration protocol follows exactly the same steps as the method described for PLT (*Protocol 4*).

Examples of freeze-substituted mouse small intestine and a single-cell suspension are illustrated in *Figures 4* and *5*.

An additional processing method for rapidly frozen tissue that should not be overlooked is the replacement of ice in frozen samples not by substitution but by molecular distillation (12). The method resembles a sophisticated

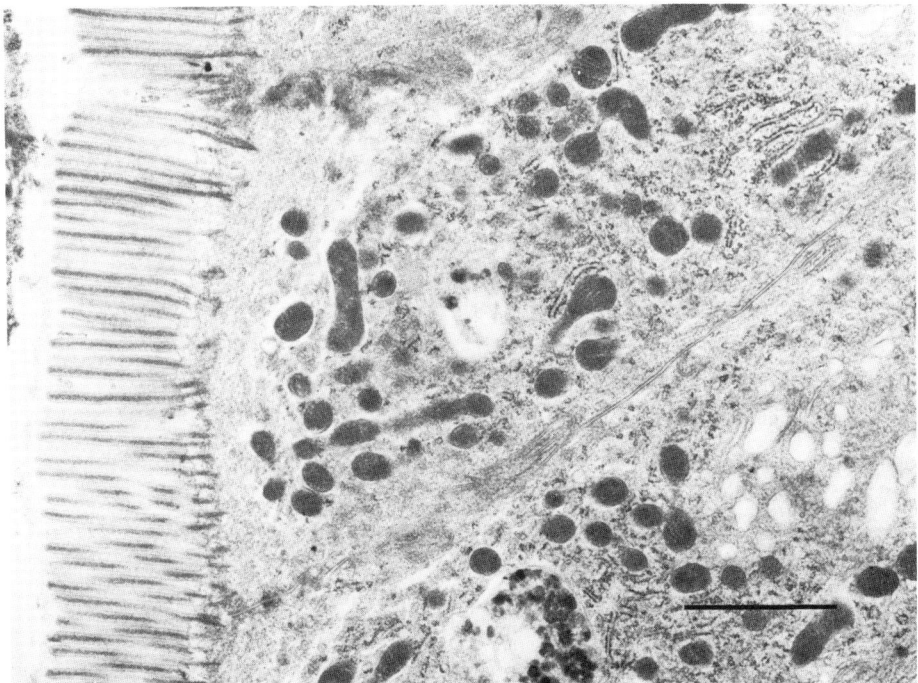

Figure 4. Mouse intestine impact-frozen, freeze-substituted in methanol, and embedded in Lowicryl HM20. Sections were contrasted with uranyl acetate and lead citrate. The impact surface was 2 μm from the microvilli surface. Scale bar 1.5 μm.

variant of freeze-drying and the protocol followed by the equipment has been devised to remove the amorphous ice phase before subsequent warming causes devitrification. Whilst good results have been produced by this system, the apparatus is expensive. It has recently been configured to make advantage of the Lowicryl resins for low temperature embedding and it will be interesting to see the results of this method.

3.4 Immunocytochemistry of resin-embedded material

The immunocytochemical procedures for both PLT and freeze-substituted material are the same.

With any new reagent or tissue, it is much easier in the long run to determine the antibody dilutions and indeed the localization of the antigen at the light microscope level before proceeding to ultrathin sections.

3.4.1 Sectioning Lowicryl resins

Well polymerized blocks of HM20 will section easily, and behave much like Epon blocks. If the resin is not fully polymerized, sectioning difficulties will occur, and in addition, the sections may break up during immunolabelling.

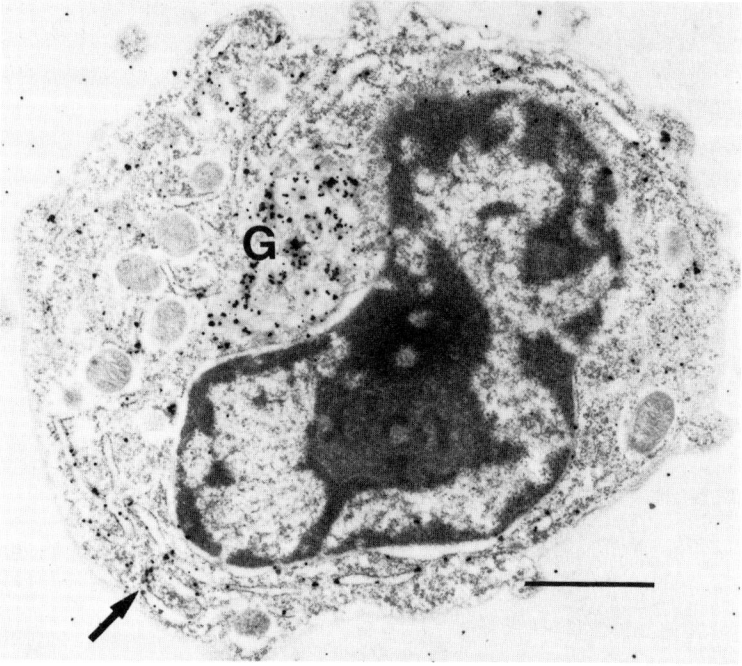

Figure 5. Blast cell from sheep lymph impact-frozen and freeze-substituted in methanol. Sections were immunolabelled with a polyclonal rabbit anti-sheep Fab′ followed by 1 nm gold conjugate silver enhanced for 12 min. The sheep immunoglobulins are predominantly seen in the Golgi (G) and rough endoplasmic reticulum (arrow). Scale bar 1.5 μm.

HM20 blocks are more brittle than epoxy resin blocks and therefore good sectioning requires that the block is carefully trimmed and the block face is free from chatter and well faced.

3.4.2 LM immunocytochemistry on resin sections

For preliminary experiments, blocks should be sectioned on to a trough filled with distilled water. This step is simplified by using a 'Histodiamond' produced by Diatome (Switzerland). Floating sections may be expanded by chloroform vapour and collected on to a subbed microscope slide. If the block face is rough, or the slides are not prepared to assist section adherence to the glass as described below, the sections wash off during the staining steps.

Several sections are placed onto each slide and with care a series of antibody dilutions can made on each slide. Each section on the slide must be surrounded by a hydrophobic ring (e.g. Dako Pen for immunohistochemistry, Denmark, UK) to prevent mixing of reagents, or alternatively a separate slide is used for each section.

The protocols that follow for both light and electron microscope immuno-cytochemistry should be regarded as a good starting point. If they do not work perfectly with a particular combination of resin/reagents, they may require modification. Prepare the following solutions:

- subbing solution (*Protocol 6*)
- blocking buffer (see below)
- washing buffer (see below)
- serial dilutions of primary antibody in blocking buffer
- 5 nm gold conjugate at appropriate dilution of blocking buffer
- ultrapure distilled water (18 MΩ); lower quality water will introduce background at the silver enhancement stage
- silver enhancement kit or reagents for the Danscher method (13)

Protocol 6. Preparation of subbed slides

Subbed slides can be prepared and stored for use. Make up 200 ml of subbing solution consisting of 1% gelatin plus 1% formalin in distilled water.

1. Add 2 g gelatin to 194 ml distilled water.

2. Warm to dissolve.

3. Add 6 ml of 36% formalin.

4. Immerse microscope slides in subbing solution, drain, and dry in a 60 °C oven.

Blocking buffer and washing buffer can be prepared as follows:

(a) Blocking buffer: PBS, plus 0.8% BSA, plus 0.1% gelatin, plus 5% foetal calf serum (FCS) at pH7.4

(b) Washing buffer: PBS, plus 0.8% BSA, plus 0.1% gelatin at pH7.4

Protocol 7. LM immunocytochemistry of resin sections

1. Cut 0.5 μm sections on to a water-filled trough.

2. Transfer to subbed slides and allow to dry.

3. Incubate with blocking buffer for 30 min.

4. Incubate with primary antibody at room temperature in a moist chamber overnight.

5. Wash three times in washing buffer, 5 min each wash.

6. Add second antibody- or protein A-5 nm gold conjugate at a suitable dilution.

Protocol 7. *Continued*

7. Incubate at room temperature for 90 min.
8. Wash three times in washing buffer, 5 min each wash.
9. Wash three times in ultrapure water, 2 min each wash.
10. Incubate with silver enhancer—the timing will have to be determined for the reagents in use. For Amersham (UK, USA) Intense M kit, 20 min at 20 °C is the correct timing (see also Chapter 3, *Protocol 11*).
11. Wash three times in ultrapure water, 2 min each wash.
12. Stain for 1 min with 0.1% Toluidine Blue.
13. Wash in distilled water and dry.

Using this method both the primary and secondary antibody dilutions can be established to give clean specific staining. Suitable controls should always be included at this stage.

3.4.3 Immunocytochemical controls

For polyclonal antibodies, an adsorption control is the best control to use. If purified antigen is not available, replacement of the primary antibody with normal serum of the appropriate animal is the next best option. For monoclonal antibodies, the primary antibody should be replaced with an inappropriate antibody of the same subclass.

3.4.4 EM immunocytochemistry on resin sections

Prepare:

- Formvar (or similar)-coated gold or nickel grids
- blocking buffer (see Section 3.4.2)
- washing buffer
- primary antibody
- 5 nm gold conjugate
- silver enhancer (optional)

Cut ultrathin sections on to Formvar (or similar)-coated gold or nickel grids. Gold grids although more expensive are easier to manipulate. Lowicryl sections tend to be less robust than epoxy resin sections and may need the support of the Formvar film during the labelling and washing steps. This does, however, mean that only one side of the section is available for immunolabelling. For many antigens this will not be a problem, but for less abundant antigens, labelling both sides of the section will double the sensitivity of labelling. Only if the resin block is well infiltrated and polymerized will the sections be able to withstand the labelling procedures without a Formvar support. For this double-sided labelling if silver enhancement is used the gold grids will prove a problem as the grids themselves will be enhanced. Nickel grids should therefore be chosen for this method. For single-sided labelling, if the sections have been collected on the Formvar side of the grid, this side will be floated on the reagents and silver enhancement will not be a problem. If the grid sinks, however, silver enhancement cannot be used.

Having established the reagent dilutions by a series of light microscope labelling experiments, the same conditions are used for the electron microscope sections. The only difference is that the silver enhancement incubation is drastically reduced or omitted.

What is the optimal size of gold colloid marker? Given that the larger (>15nm) gold colloid, whilst easy to detect, will give markedly reduced labelling intensity in comparison to smaller probe sizes, 10 nm gold is a good compromise to use for abundant antigens. There is, however, significant improvement in labelling efficiency by the use of 5 nm gold conjugates. It is our routine to use 5 nm gold conjugates for resin sections and where the small size makes identification difficult, to silver enhance the 5nm gold for 6 min. The silver enhancement process is temperature dependent and it is important to ensure that the time/temperature relationship is controlled. Different commercial kits for silver enhancement will also require different incubation times.

Protocol 8. EM immunocytochemistry

All incubations are on droplets of buffer or reagents on a sheet of 'Fuji' film or Parafilm. Minimum droplet size is 10 μl, but care must be taken to ensure these small drops do not evaporate. Transfer grids with forceps or an elongated wire loop. The elongated loop discourages the transfer of reagent between droplets. Depending on the conditions described above the grids are either floated on the surface of the reagents or sunk, in the latter case, transfer of grids is by forceps.

1. Block non-specific binding with blocking buffer for 30 min.
2. Incubate on droplets of antibody diluted in blocking buffer, overnight.
3. Wash three times with washing buffer, 5 min each wash.
4. Incubate in 5 nm gold conjugate diluted in blocking buffer for 90 min.
5. Wash three times with washing buffer, 5 min each wash.
6. Silver enhancement is optional at this stage.

Wash in distilled water and contrast if no enhancement is needed.

For silver enhancement
7. Wash three times in ultrapure water, 2 min each wash.

Protocol 8. *Continued*

  8. Silver enhance for the required time.
  9. Wash three times in ultrapure water, 5 min each wash.
  10. Contrast.

The preferred method of contrasting needs to be established, but in this laboratory the Ultrostainer (Leica, UK) is used with the same settings as for routine epoxy resin sections.

4. Cryosection techniques

The previous text has described methods where the immunological steps are performed on sections of resin-embedded material. PLT embedding is appropriate for antigens withstanding a short fixation regime and freeze-substitution can be undertaken in the absence of fixatives. On the surface, therefore, it would seem that these two methods cover almost all eventualities. The one major situation not covered is where the antigen is damaged by the dehydration/substitution solvents or the resin monomers. In this case, unless one of the less rigorous dehydration protocols which can be applied to LR White embedding can be used, no resin embedding method is indicated Cryosectioning followed by thawing and immunolabelling of the sections is the other choice. It is our experience that there are antigens which are not detectable in resin-embedded material but which are nonetheless readily demonstrated in cryosections. An example of such an antigen is illustrated in *Figure 6.*

Immunolabelling of resin sections takes place only at the surface. It might be considered that an additional advantage of using thawed cryosections would be an increase in sensitivity. As will be discussed later, this is regrettably not the case.

4.1 The preparation of thawed ultrathin cryosections

To avoid confusion with frozen sections which are maintained in the frozen state within the electron microscope, these sections should more accurately be called 'thawed cryosections'. The method owes much to the work of Tokuyasu who has established many of the key parameters for successful immunolabelling of thawed cryosections (14).

Unlike the two methods described for resin embedding (PLT and freeze-substitution), where construction of suitable apparatus is a viable option, the construction of a cryomicrotome would be beyond almost any laboratory. Cryo-attachments are available for three routine ultramicrotomes; two from Leica (UK) and one from RMC (UK).

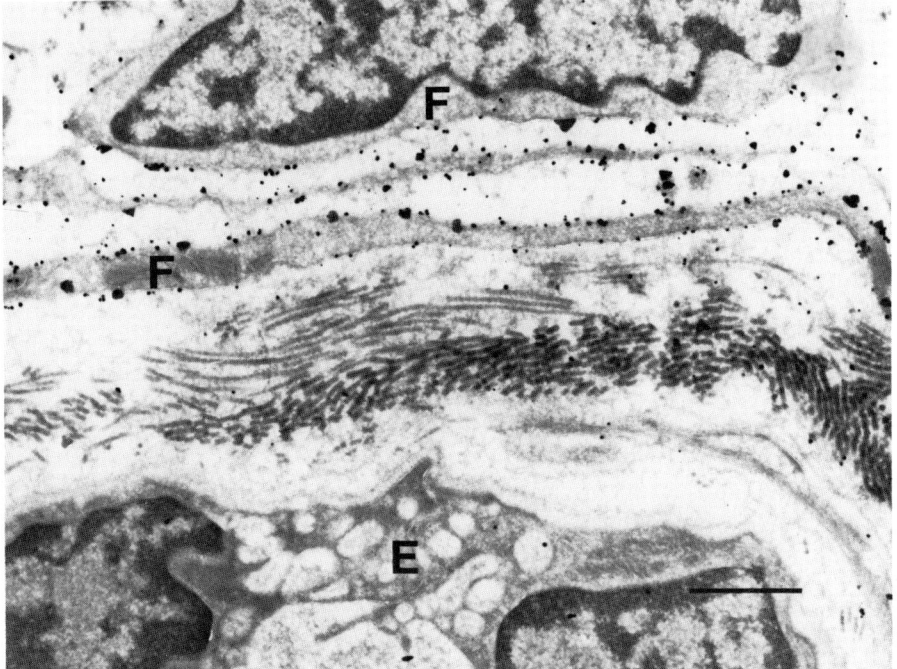

Figure 6. Cryosection of human breast. The sample was fixed for 1 h in 2% paraform-aldehyde plus 0.05% glutaraldehyde fixative. The section was immunolabelled with an anti-CD13 antibody followed by 1 nm gold conjugate, silver enhanced for 9 min. The capillary endothelial cell (E) is CD13 negative, whilst the fibroblasts (F) express CD13 on their membranes. Scale bar 1 μm.

The main characteristics of cryosections as used for immunocytochemistry are:

(a) advantages:
- no dehydration and embedding
- some membraneous organelles well preserved

(b) disadvantages:
- some organelles difficult to detect
- non-permanent blocks
- aldehyde fixation required
- specialized microtome needed

Cryosectioning for immunocytochemistry consists of the following steps:

(a) brief aldehyde fixation
(b) cryoprotection

(c) freezing

(d) sectioning

(e) thawing and immunolabelling sections

(f) contrasting and embedding sections

4.1.1 Fixation

The same constraints upon fixative constituents and timing that have been described for PLT embedding apply to cryosections. For cells from tissue culture, or samples in suspension, some support matrix is needed to maintain the integrity of the sections of the cell suspension after thawing.

Protocol 9. Fixation and embedding of cell suspension samples for cryosectioning.

1. Fix samples in 2.05 as a suspension for 60 min.
2. Centrifuge at 500g for 5 min. (This speed should be varied to suit other types of specimen).
3. Re-suspend in PBS.
4. Centrifuge at 500g for 5 min (or other speed to suit particular specimens).
5. Re-suspend in 2% gelatin in PBS at room temperature and leave for 60 min.
6. Centrifuge at 500g for 5 min (or other speed to suit particular specimens).
7. A secondary fixation (as step **1**) is optional at this stage.
8. Cool on ice and remove from the centrifuge tube (plastic centrifuge tubes will assist this as the bottom can be cut off).
9. Keep the gelatin on ice and cut small pieces of gelatin containing the sample. These should be small (> 0.5 mm cube) and if possible pyramidal.
10. Transfer immediately to cryoprotectant.

4.1.2 Cryoprotection

This is required to prevent the formation of ice crystals during freezing. Samples which have been infiltrated with cryoprotectant can be frozen relatively slowly without risk of freezing damage. Dropping the samples into liquid nitrogen is an adequate freezing method and the whole of the block will be well frozen.

Wash the sample in PBS for 5 min after fixation and transfer to a solution of 2.3 M sucrose in phosphate buffer. The sample is fully infused when it is completely translucent. This will take from 30 min to overnight, but should not be prolonged more than necessary.

The cutting properties of cryoprotected samples are determined to a considerable extent by the sectioning temperature and by the nature of the cryoprotectant. Mixtures of sucrose and either PVP or PVA as the cryoprotectant have been recommended as giving superior cutting properties to 2.3 M sucrose (15).

4.1.3 Freezing

Select well trimmed pieces of tissue or gelatin containing the sample which have been cryoprotected and place in a drop of 2.3 M sucrose (or sucrose PVP/PVA mixture) on the specimen holder. Centralize the block, and if pyramidal ensure the base is on the specimen holder. With a small piece of filter paper, remove most of the sucrose, leaving the base of the block sitting in a small amount of 2.3 M sucrose to act as 'glue'. Freeze the specimen by dropping the pin into LN_2, and either transfer to LN_2 for storage or introduce directly into the cryoultramicrotome chamber.

4.1.4 Trimming blocks

Transfer the specimen pin to the cooled chamber of the microtome with the temperature set at −80 °C. Sectioning will be at cooler temperatures, but below −90 °C the blocks become brittle and difficult to trim.

Before trimming for thin sections, it is wise to ensure that the region of interest in the block is at the cutting face. Carefully trim away the front edge of the block on the trimming tool or a glass knife. Initially the trimmings will be 2.3 M sucrose, but as the block face becomes larger the specimen will be exposed and the trimmings will change from white and friable to recognizable sections. Cut several 0.5 μm sections and move them away from the knife edge with a human eyelash attached to a wooden or plastic handle. If using an anti-static device, switch off at this stage to prevent the sections flying away. The sections are collected upon a droplet of 2.3 M sucrose held in a small-diameter (approximately 2.5 mm) wire loop. Dip the loop into the sucrose to collect a drop and move it into the cryochamber. It will take several seconds to freeze. When it stops 'smoking', but before it freezes, touch the drop on to the sections and withdraw from the chamber. This step is facilitated by the sucrose droplet forming a lens allowing the sections to be seen through the droplet.

Wait until the droplet warms (speeded by holding close to a 'cupped' hand) and place the drop on to a microscope slide. The sections on the lower surface of the drop will adhere to the microscope slide. Wash carefully with distilled water and stain with dilute (0.1% Toluidine Blue) for 30 sec. Wash with distilled water, allow to dry, and check with a light microscope. If the region of interest is present, the final trimming can be done. As with resin sectioning, the quality of the sectioning depends crucially upon the trimming

Figure 7. The chamber of a Leica FC4 cryokit. The knife holder is shown with glass and diamond knives, and between them is the Diatome trimming diamond (arrow). The position of the anti-static probe (P) is also shown.

of the block. This can be done in several ways; either by using the lateral edges of glass knives, a metal trimming tool, or by using a Diatome diamond trimming tool. This latter device greatly assists block preparation. The facing of the block, cutting of thick sections to assess the presence of tissue within the block, and the final trimming of the block can all take place using this one tool (*Figure 7*).

4.1.5 Glass or diamond knives

A diamond knife does not give the same improvement over glass knives as is seen with resin sectioning. Thus for the majority of work, a glass knife is adequate.

i. Glass knives

If glass knives are to be used, it is important that they are of the highest quality. The knife angle should be as near to 45 ° as possible. This can be determined by assessing the counterpiece of the complementary knife. (This is the foot of the other knife broken from the square.) The foot should be as thin as possible and of even thickness. Ideally the foot should be in the region of 0.1 mm in width, but in practice these knives are not always easy to prepare and a marginally thicker foot may suffice. Knives should be made by the balanced break method where each break is at the centre of the glass strip. The strip is broken in half and half again until squares are formed. These are broken into the knives. The final break should be a slow break, and this is easily achieved by applying pressure until the score begins to show signs of extending through the glass and leaving the knife to break at its own speed.

ii. Diamond knives

The cutting properties of diamond knives have recently been studied by H. Gnaegi of Diatome and several findings have emerged. Two parameters in particular affect the quality of sectioning. These are the effect of static as the sections are cut and the appearance of mechanical compression of the sections. The effects of the various cutting parameters on the section quality are summarized in *Table 3*.

Table 3. Summary of the effects of altering cutting conditions on the quality of cryo-sections

		Static	Mechanical compression
Glass knives (compared with diamond knives)		Less	Medium
Diamond knives	(45 °)	Medium	Medium
	(35 °)	More	Less
Sectioning temperature	(warmer)	Less	More
	(colder)	More	Less
Sectioning speed	(> 5 mm/sec)	No change	More
	(< 5 mm/sec)	No change	Less

In short, the reason why diamonds have been disappointing in cryosectioning is that they are affected more than glass knives by static, and this tends to stick the section to the knife resulting in compression. To overcome the static problem an anti-static device has been developed and is available from Diatome (Switzerland). Whilst its benefits are best observed with diamond knives, the device also improves sectioning with glass knives. The use of the device requires care in positioning the probe, but the improvement in the cutting abilities of a diamond knife is considerable. If the probe is too close to the cutting edge the ribbon of sections tends to fly away, and if placed too far away the beneficial effects of the device are diminished. Using the trimming tool and a diamond knife with the anti-static system in the Leica FC4 (UK), the ease and speed of cutting cryosections is much improved (see *Figure 7*).

4.1.6 Sectioning temperature

Different samples will cut best at different temperatures and so it is difficult to give exact figures. The range is between −90 and −110 °C. In general, samples embedded in gelatin will cut better at slightly warmer temperatures than solid tissues. The optimum sectioning temperature for each sample can only be finally determined by trial and error.

4.1.7 Sectioning speed

The best sections are cut at slow speeds. Optimally this will be in the 0.5–5 mm/sec range. Only when the block is adequately faced will it section well and for glass knives it may assist sectioning to cut at a high speed (10 mm/sec) until sections are ribboning, and then slowly reduce the cutting speed. This latter approach will not, however, benefit sectioning when using an anti-static device. In this case, sectioning should begin at around 0.5–1 mm/sec.

Sections are collected as for thick sections described above, and the thawed sections are placed on to carbon/Formvar-coated gold grids. The grids can be floated, section side down, on droplets of PBS or placed section side down on a layer of 2% gelatin in PBS in the bottom of a petri dish. Sections do not last indefinitely, and should not be kept more than 48 h.

4.2 Immunocytochemistry on cryosections

The steps to immunolabelling cryosections are similar to *Protocol 8* and are as follows:

(a) block non-specific labelling

(b) primary antibody

(c) wash

(d) gold conjugate

(e) wash

(f) silver enhance if required

(g) contrast and embed

The blocking procedure seems to vary with each laboratory. For cryosections, 5% FCS in PBS has proved adequate for many of our studies and is used routinely. Primary antibody dilutions in FCS/PBS will be similar to when the reagent is used for resin sections, but if this information is not available, it will have to be determined by previous experiments at the LM level.

In the initial stages of the development of cryosections for immunocyto-chemistry, it was believed that the whole thickness of the section would be available for immunolabelling and that consequently a huge increase in sensitivity over resin sections could be expected. When this improvement did not materialize, it was clear that the reagents, and the gold conjugate in particular, did not penetrate the sections (11). Ultrasmall (1 nm) gold conjugates were therefore developed to try to aid the penetration of the sections. These 1 nm conjugates can, in many instances, give a useful increase in immunolabelling over 5 nm conjugates in cryosections (17). Their small size does, however, mean that a silver enhancement step is necessary to detect the gold marker. Interestingly, the increase in immunolabelling observed on cryosections when using 1 nm gold conjugates has not been apparent in immunolabelling experiments on resin sections (Robertson, unpublished data). This observation does add weight to the concept that these small markers do have improved accessibility to antigens in cryosections, but not in resin sections.

4.2.1 Contrasting cryosections

A number of methods have been described for contrasting cryosections. The simplest method is described below. More complex methods are described elsewhere (18).

Prepare

- 3% solution of uranyl acetate in ultrapure water
- 2% solution of methyl cellulose in ultrapure water

Protocol 10. Preparation of methyl cellulose for embedding cryo-sections

1. Warm 100 ml ultrapure water to 95 °C.
2. Add 2.0 g methyl cellulose (viscosity 25 cP).
3. Stir for 5 min.
4. Seal container and transfer to 4 °C.
5. Stir for 3–4 days.
6. Centrifuge at 60 000*g* for 60 min.
7. Remove the supernatant for storage at 4 °C, ready for use.

Protocol 11. Contrasting and embedding cryosections

1. Mix 100 µl of 3% uranyl acetate with 900 µl of the methyl cellulose prepared in *Protocol 10*.
2. Place droplets on to wax or 'Fuji' film on ice.
3. Transfer grids of cryosections from ultrapure water at the end of immunolabelling onto droplets of uranyl acetate/methyl cellulose for 1 min.
4. Transfer to fresh droplets of uranyl acetate/methyl cellulose.
5. Cover and incubate for 15 min.
6. Make a wire loop, 4 mm in diameter from 25 µm copper wire and dip the loop under the grid and pick up the grid which will be held in the centre of the loop by a film of methyl cellulose.

Protocol 11. *Continued*

7. Hold the loop vertically and touch on to a filter paper. When the methyl cellulose wets the filter paper, run the loop over the paper to remove excess methyl cellulose. The grid should still be held in a film of methyl cellulose.

8. Allow to dry for 15–20 min.

9. Cut round the edge of the dried methyl cellulose film with a sharp implement to remove the grid (the tip of a syringe needle is ideal) for examination with the electron microscope.

The object of this uranyl acetate/methyl cellulose stage is two fold. Firstly, to provide some contrast to the sections, and secondly to give the sections some support during drying. The thickness of the methyl cellulose film is crucial. Too thick and the electron beam will not penetrate; too thin and the morphology of the sections suffers. The thickness of the copper wire and the diameter of the loops will both affect the final thickness of the methyl cellulose embedding.

In general, however, cryosections do not give exactly the same appearance as routine resin-embedded sections. A more complex method has been described of following immunolabelling with osmium tetroxide, dehydration, and embedding in a very thin resin support (19). Although time consuming, the results are impressive and resemble resin-processed material.

5. Conclusions

The ultrastructural localization of abundant, robust antigens is relatively straightforward in that almost any processing method will provide material that can be immunolabelled. Unfortunately, the majority of antigens are present in relatively low concentrations, and are often degraded by the various stages in processing. Successful localization requires matching the antigen and the antibody combination with a compatible processing method. The methods described here, in conjunction with pre-embedding labelling methods (Sections 6 and 7), will allow almost all antigens to be detected.

Ultrastructural immunocytochemical methods are undergoing a continuous process of change and improvement and it will be many years before it will be accurate to say that these methods have reached their final form.

Acknowledgements

The Institute of Cancer Research is supported by funds from the Medical Research Council and the Cancer Research Campaign.

B. The pre-embedding techniques
Julian E. Beesley

6. Introduction

The pre-embedding techniques are used for immunolabelling outer membrane antigens on prokaryotic and eukaryotic cells (20) by incubating cell suspensions with the immunolabelling reagents (*Protocol 12* and *Figure 8*) and for localizing internal antigens (*Protocol 13* and *Figure 9*) by immuno-labelling 20–100 μm sections of the specimen (21). The immunolabelled specimen is fixed, dehydrated, embedded in an epoxy resin, and stained ultrathin sections are prepared for examination. The technique is highly sensitive because fixation before immunolabelling is not always necessary. If the specimen needs to be stored before immunolabelling and the antigen can

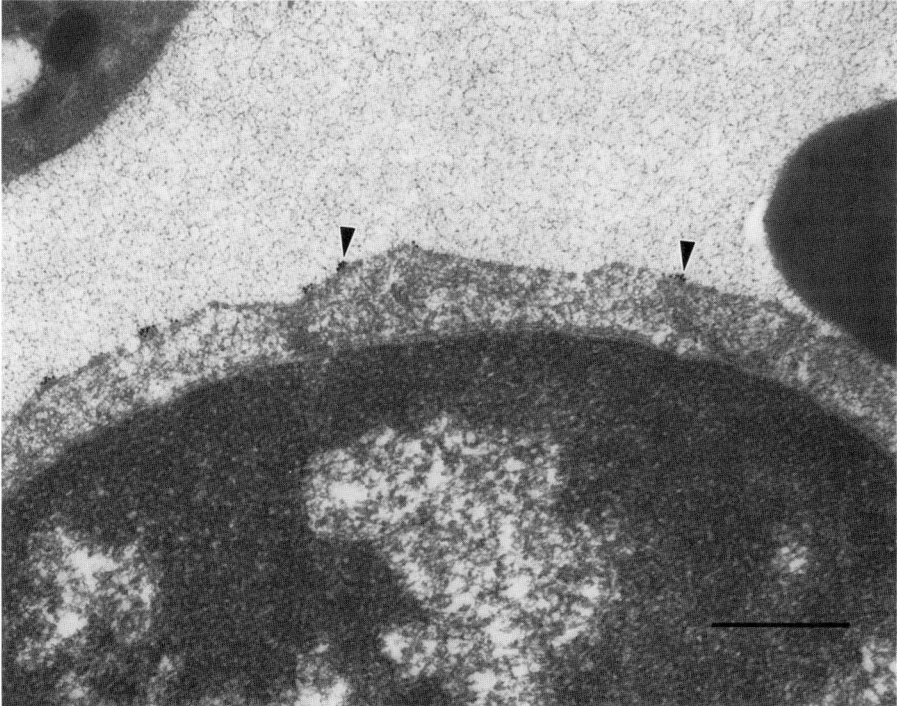

Figure 8. The pre-embedding technique (*Protocol 12*) was used to identify CD3-positive lymphocytes. The 5 nm gold probes (arrows) were localized on T lymphocytes and neither the granulocytes (top left) nor erythrocytes (top right) were immunolablled. (From a study in collaboration with Dr. J. Rhodes and Mrs A. Vardill, Welcome Research Laboratories, Beckenham, Kent.) Scale bar 0.5 μm.

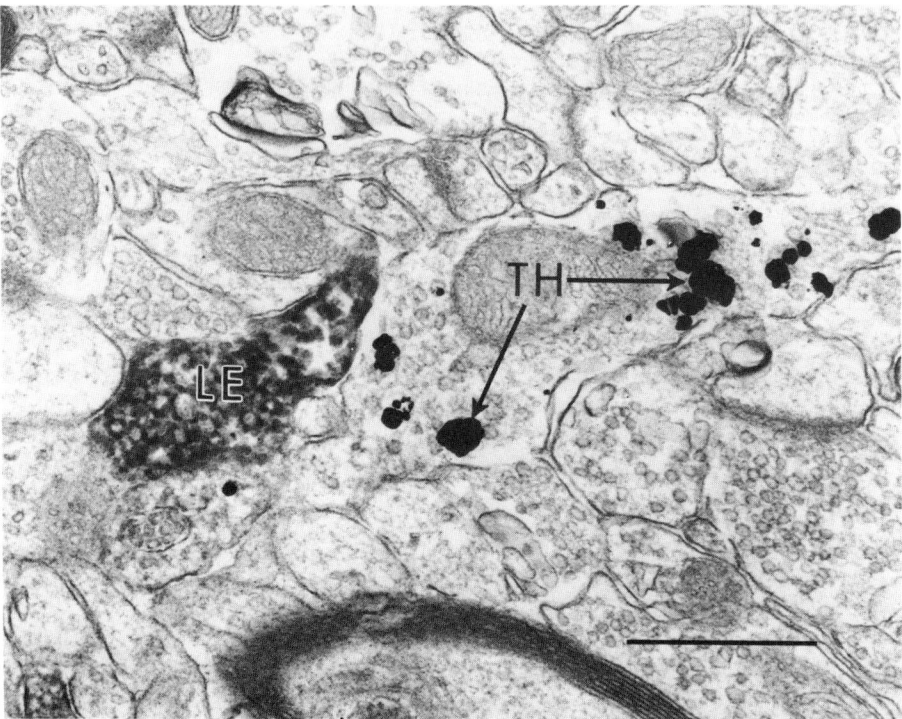

Figure 9. The pre-embedding technique (*Protocol 13*) was used for identifying two internal antigens. Immunolabelling was sequential as described in *Protocol 3*, Chapter 6, and shows dual immunogold–silver and immunoperoxidase labelling in the rat caudate nucleus. Leucine-enkephalin (LE) is localized in axon terminals containing peroxidase reaction product. Silver particles indicative of tyrosine hydroxylase (TH) labelling are also observed in axon terminals. Silver enhancement time was 6 min at room temperature. (Kindly supplied by Drs J. Chan, V. Pickel, and S. Sesack, Department of Neurology and Neuroscience, Division of Neurobiology, The New York Hospital–Cornell Medical Centre, New York, NY 10021, USA.) Scale bar 0.5 μm.

withstand aldehyde fixation, the specimen is lightly fixed and stored in buffer. Immunolabelling outer membrane antigens of eukaryotic cells is usually performed at 4 °C to prevent patching and capping (Chapter 2, Section 3).

Any antibody and electron dense probe may be used for pre-embedding immunolabelling. For immunolabelling outer membrane antigens colloidal gold is the preferred choice. The 10 nm probe is recommended for general use. Small (5 nm) gold probes are used if intense immunolabelling is desired or the examination will be at relatively high magnifications. One nanometre gold probes have recently been used for immunolabelling thick sections (*Protocol 13*).

Both single and multiple immunolabelling are possible (see Chapter 6).

7. Immunolabelling techniques

Protocol 12. The pre-embedding technique for immunolabelling outer membrane antigens

All reagents are diluted with phosphate (0.01 M, pH 7.2)-buffered saline (0.15 M) containing 1% bovine serum albumin (PBSA). If isolated cells are being immunolabelled the preparations should be agitated thoroughly every 15 min during immunolabelling to prevent bivalent antibodies clumping the cells into which the reagents, especially colloidal gold, do not penetrate. Likewise washing the cells is performed by centrifugation which should be the minimum force necessary to pellet the cells.

1. Incubate unfixed cells with at least three times their volume of specific antibody diluted with PBSA for 1 h, at 4 °C, with frequent vigorous agitation.

2. Wash cells at 4 °C, by minimal centrifugation, four times with PBSA.

3. Incubate cells with appropriate colloidal gold probe diluted with PBSA for 1 hr at 4 °C with frequent vigorous agitation.

4. Wash cells at 4 °C, by minimal centrifugation, four times with PBSA.

5. Wash cells twice in PBS by minimal centrifugation at 4 °C.

6. Fix cells with 1% buffered glutaraldehyde for 15 min at room temperature.

7. Centrifuge to a pellet and wash the pellet in buffer. If the pellet disintegrates the cells are encapsulated in 5% gelatin before embedding in epoxy resin. The embedding steps usually consist of post-fixation with 1% aqueous osmium tetroxide for 1 h, rinse (5 min) in water, dehydrate in an ascending series of ethanol, and, using propylene oxide as an intermediate reagent, embed in an epoxy resin. Ultrathin sections are stained with alcoholic uranyl acetate and acidic lead citrate before examination.

Protocol 13. The pre-embedding technique for immunolabelling intracellular antigens

Permeabilization of the tissue may be achieved by the freezing and thawing process or it may be accomplished using 0.05% saponin (22, 23) or Photo Flo (24). Alternatively the very small 1 nm gold particles are used for immunolabelling because they penetrate the tissue. They have been used in conjunction with the silver enhancement technique (25) in which the tissue is lightly fixed and thick sections are immunolabelled without the use of freeze–thawing or other methods that enhance penetration but damage ultrastructure.

Protocol 13. *Continued*

1. Fix the tissue with 3.75% acrolein and 2% freshly prepared paraformaldehyde in 0.1 M phosphate buffer for 6 min, and transfer to 2% freshly prepared paraformaldehyde in 0.1 M phosphate buffer for 1 h.

2. Wash in phosphate buffer before cutting 30–40 µm sections with a vibratome.

3. Treat with 1% sodium borohydride in 0.1 M phosphate buffer for 30 min.

4. Wash thoroughly in phosphate buffer.

5. Rinse thoroughly with 0.1 M Tris-buffered saline, pH 7.6 (TBS).

6. Incubate for 18–24 h in a predetermined optimal dilution of antibody in TBS.

7. Wash three times, 5 min each, in 0.1 M phosphate-buffered saline (0.15 M), pH 7.4 (PBS).

8. Incubate sections for 5 min with 0.2% gelatin in PBS containing 0.8% bovine serum albumin.

9. Incubate with a suitable 1 nm gold probe diluted with 0.1% gelatin in PBS with 0.8% bovine serum albumin for 3 h.

10. Wash three times with PBS, 5 min each wash.

11. Rinse briefly (three times, 1 min each) with 0.2 M citrate buffer, pH 7.4.

12. Silver enhance (see Chapter 3).

13. Terminate the silver enhancement by two 1 min washes with 0.2 M citrate buffer, pH 7.4.

14. Rinse twice for 1 min each in 0.1 M phosphate buffer.

15. Fix for 30 min with 2% osmium tetroxide in 0.1 M phosphate buffer.

16. Wash with 0.1 M phosphate buffer for 5 min, dehydrate in an ascending series of ethanol, and embed in an epoxy resin.

17. Collect ultrathin sections from the outer regions of the specimen for examination.

C. The immunonegative stain technique
Julian E. Beesley

8. Introduction

Immunolabelling small specimens such as bacterial pili, viruses, or isolated cell organelles poses several problems. They are so small that it is difficult to prepare a pellet which can be fixed, dehydrated, and embedded for the preparation of ultrathin sections for immunolabelling. If sections were prepared, the thickness of the resin layer, albeit 80 nm, would inhibit immunolabelling of particles such as viruses which are commonly less than 100 nm in diameter. The immunonegative stain technique was developed specifically to overcome these problems (26). The specimen is dried on to a coated electron microscope grid and immunolabelled with antibody and electron dense probe on the grid. The specimen is negatively stained which produces a high resolution image of the specimen for detailed observation (*Figure 10*). The immunonegative stain technique localizes external antigens only. Fixation is usually not necessary and the efficiency of immunolabelling is therefore high. Any specific antibody can be used, monoclonal or polyclonal, in conjunction with the appropriate colloidal gold probe. The specimens are usually small and therefore either 5 or 10 nm probes are used.

Since the preparation of the specimen on the grid needs so little sample

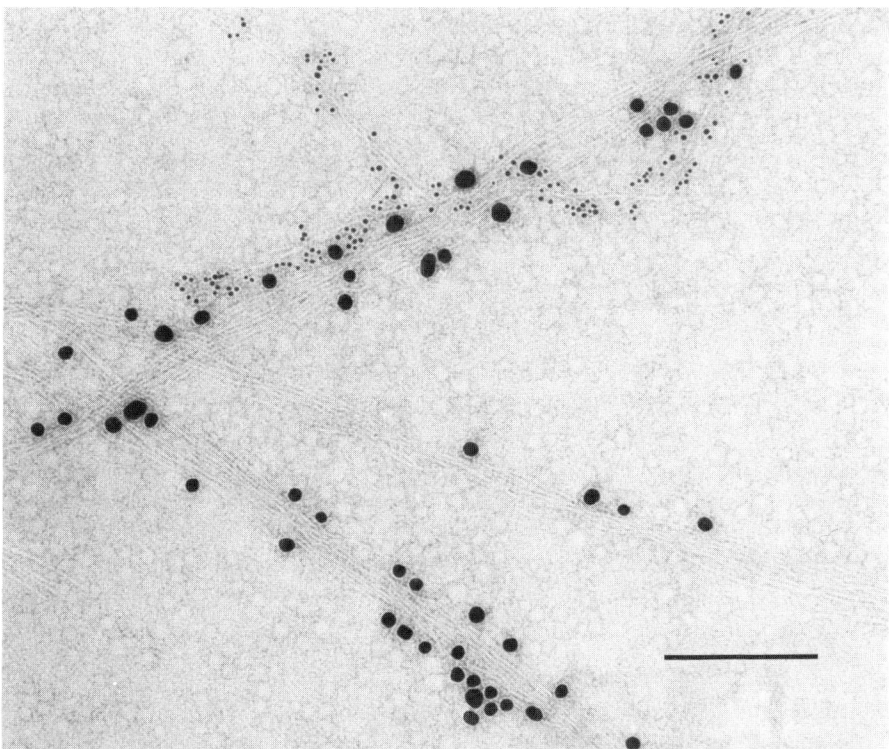

Figure 10. The immunonegative stain technique (*Protocols 14* and *15*) for identifying bacterial pili serotypes. Pili were dried on to a grid and incubated sequentially with first primary antibody, a 5 nm Protein A–gold probe, the second antibody and finally a 15 nm Protein A–gold probe before contrasting. The concentrations of the reagents were adjusted so that there were no cross reactions. Scale bar 0.2 μm.

(5 μl), several grids can be prepared from a very small amount of starting material, commonly a centrifuged pellet, and subsequently immunolabelled with different antibodies or different concentrations of antibodies.

The technique is adaptable and has been applied to eukaryotic cells grown on coated electron microscope grids and immunolabelled *in situ* (27, 28).

9. Immunolabelling technique

Immunolabelling is performed by preparing a series of unstained specimens on grids (*Protocol 14*) and immunolabelling them with the required antibodies and probes (*Protocol 15*).

Protocol 14. Preparation of the specimen for immunolabelling

The objective of this step is to prepare a concentrated suspension of the sample and dry it on to a coated electron microscope grid so that the density is sufficient for observation after immunolabelling but not too concentrated so that the immunolabelling of one organism interferes with that of another making interpretation difficult.

1. Centrifuge the specimen to a pellet, preferably in distilled water or alternatively phosphate (0.1 M, pH 7.2)-buffered 0.15 M saline (PBS). The centrifugation speed will depend on the specimen. For viruses 15 000g for 1 h is recommended, but for bacteria 3000g for 5 min is sufficient.

2. Carefully remove the supernatant and re-suspend the specimen in distilled water or PBS until a milky suspension is obtained.

3. Place a 15 μl droplet of the suspension on to a sheet of Parafilm or dental wax and float a 400 mesh Butvar-coated gold grid on this for 1 min.

4. Remove the grid with a pair of tweezers but do not dry. If the sample has been prepared in PBS quickly wash the sample for 1 min on a droplet of distilled water. After this, or if the specimen had originally been prepared in distilled water, blot the grid almost dry with the edge of a piece of filter paper.

5. Float the specimen for 30 sec on a droplet of negative stain, 3% sodium phosphotungstate pH 6.8 if a virus or cell organelle or 1.5% ammonium molybdate pH 6.8 if a bacterium.

6. Remove the grid with a pair of tweezers, blot dry and examine in the electron microscope.

The specimen should be scattered evenly across the grid and be well stained. If not, adjust preparation conditions accordingly. It the specimen is too concentrated, dilute the suspension further at step **2**. If it is too dilute add less fluid at step **2**. If the staining is too dense dilute the stain further or if the staining is too weak prepare and use a more concentrated stain.

When the correct parameters have been found prepare the requisite number of grids for immunolabelling allowing for controls but **do not stain** until after immunolabelling.

Protocol 15. Immunolabelling specimens dried on to electron micro- scope grids

All immunolabelling is performed by floating the unstained grids prepared as above (*Protocol 14*) on droplets of reagents placed on a sheet of Parafilm or dental wax. All reagents are diluted with phosphate (0.01 M, pH 7.2)- buffered saline (0.15 M) containing 1% bovine serum albumin (PBSA).

1. Float an unstained specimen on a 15 µl droplet of antibody suitably diluted with PBSA for 15 min.

2. Wash four times for 1 min each wash by floating the grid specimen-side down on droplets of PBSA.

3. Float the grid, specimen-side down on a 25 µl droplet of colloidal gold probe, diluted with PBSA, for 15 min.

4. Wash four times for 1 min each wash by floating specimen-side down on distilled water.

5. Stain with either 3% sodium phosphotungstate or 1.5% ammonium molybdate pH 6.8 as in *Protocol 14* before viewing with the electron microscope.

References

1. Monaghan, P. and Roberts, J. D. B., (1985). *J. Pathol.*, **147**, 281.
2. Bendayan, M. and Zollinger, M. (1983). *J. Histochem. Cytochem.* **31**, 101.
3. Mitchell, D., Ibrahim, S., and Gusterson, B. A. (1985). *J. Histochem. Cytochem.*, **33**, 491.
4. Newman, G. R., Jasani, B., and Williams, E. D. (1983). *Histochem. J.*, **15**, 543.
5. Armbruster, B. L., Carlemalm,. E., Chiovetti, R., Garavito, R. M., Hobot, J. A., Kellenberger, E. *et al.* (1982). *J. Microsc.*, **26**, 77.
6. Kellenberger, E., Villiger, W., and Carlemalm, E., (1986). *Micron. Microsc. Acta*, **17**, 331.
7. Monaghan, P. and Robertson, D. (1990). *J. Microsc.*, **158**, 355.
8. Robards, A. W. and Sleytr, U. B. (1985). In *Practical Methods in Electron Microscopy*, Vol. 10, (ed. A. M. Glauert), pp. 461–95. Elsevier, Amsterdam.
9. Humbel, B., Marti, T., and Müller, M. (1983). *Beitr. Electronmikroskop. Direktabb. Oberfl.*, **10**, 585.

10. Ryan, K. P., Bald, W. B., Neumann, K., Simonsberger, P., Purse, D. H., and Nicholson, D. N. (1990). *J. Microsc.* **158**, 365.
11. Studer, D., Michel, M., and Müller, M. (1989). *Scanning Microsc. Suppl.* **3**, 253.
12. Linner, J. G., Livesey,. S. A., Harrison, D. S., and Steiner, A. L. (1986). *J. Histochem. Cytochem.*, **34**, 1123.
13. Danscher, G. (1981). *Histochemistry*, **71**, 81.
14. Tokuyasu, K. T. (1986). *J. Microsc.*, **143**, 139.
15. Tokuyasu, K. T. (1989). *Histochem. J.*, **21**, 163.
16. Stierhof, Y-D., Schwarz, H., and Frank, H. (1986). *J. Ultrastruct. Mol. Struct. Res.*, **97**, 187.
17. Monaghan, P. and Atherton, A. (1992). In *Electron Microscopic Immunocytochemistry*, (ed. J. M. Polak and J. V. Priestley), pp. 123–136. Oxford University Press.
18. Tokuyasu, K. T. (1978). *J. Ultrastruct. Res.*, **63**, 287.
19. Keller, G.-A., Tokuyasu, K. T., Dutton, A. H., and Singer, S. J. (1984). *Proc. Natl. Aca. Sci. USA*, **81**, 5744.
20. Beesley, J. E. (1988). *Scanning Microsc.*, **2**, 1055.
21. Priestley, J. V., Alvarez, F. J., and Averill, S. (1992). In *Electron Microscope Immunocytochemistry*, (ed. J. M. Polak and J. V. Priestley), pp. 89–121. Oxford University Press.
22. Willingham, M. C. (1983). *J. Histochem. Cytochem.*, **31**, 791.
23. Ohtsuki, I., Manzi, R. M., Palade, G. E., and Jamieson, J. D. (1978). *Biol. Cellulaire*, **31**, 119.
24. Wouterlood, F. G., Sauren, Y. M., and Pattiselanno, A. (1988). *J. Chem. Neuroanat.*, **1**, 65.
25. Chan, J., Aoki, C., and Pickel, V. M. (1990). *J. Neurosci. Methods*, **33**, 113.
26. Beesley, J. E., Day, S. E. J., Betts, M. P., and Thorley, C. M. (1984). *J. Gen. Microbiol.*, **130**, 1481.
27. Hyatt, A. D., Eaton, B. T., and Lunt, R. (1987). *J. Microsc.*, **145**, 97.
28. Hyatt, A. D. and Eaton, B. T. (1990). *Electron Microsc. Rev.*, **3**, 1.

5

Special preparation methods for immunocytochemistry of plant cells

MOHAN B. SINGH, PHILIP E. TAYLOR, and R. BRUCE KNOX

1. Introduction

Plant cells are surrounded by a thick and usually rigid glycan cell wall. Inside this wall, the plant cell is essentially similar to an animal cell in being bounded by a plasma membrane and in its nucleus and complement of organelles such as mitochondria. Plant cells differ from animal cells in possessing chloroplasts and other forms of plastids, vacuoles, surface cuticle and lipid coatings, desiccated states of seeds and pollen, presence of storage organelles, starch granules, and secondary compounds such as phenolics.

These differences have meant that special preparation methods are needed, as well as special considerations in making and using antibodies. This chapter is a review of techniques, and provides current, tested protocols for plant immunocytochemistry from our own laboratory and from the recent literature. Many introductions to the field have been written (1–4) and the techniques for immunofluorescence are well established. Here, we consider plant tissue preparation for use of immunogold labelling and viewing by transmission electron microscopy. This technology provides the greatest information and resolution that is currently available.

2. Nature of antigens

2.1 Polyclonal and monoclonal antibodies

Both polyclonal and monoclonal antibodies have been employed for immunochemistry of plant cells. Polyclonal serum is a mixture of a family of antibodies to different epitopes in the antigen. Some of these antibodies may be directed against carbohydrate moieties present in a glycoprotein antigen with consequent loss of specificity (see below).

Monoclonal antibodies, because they are directed against a single epitope, may not show a sufficiently strong signal in immunocytochemical reactions. This is especially the case with less abundant antigens. Mixed monoclonal antibodies, which are directed against different epitopes of the same antigen, may provide a solution.

2.2 Protein antigens

If the antigen that is to be located in cells or tissues is abundant, it can be purified to almost homogeneity by standard techniques of protein biochemistry. Difficulties in the preparation of polyclonal antibodies arise if the protein antigen is not abundant. In these cases some of the antibodies produced in the serum may be directed against impurities in the antigen preparation. Methods have been devised to affinity purify monospecific antibodies from such antisera. This is achieved by incubating Western blots of the tissue proteins with the crude antiserum, excising the band corresponding to the antigen required and eluting off the specific antibodies at acidic pH (5). Very small amounts of antibodies are obtained in this way, and it is preferable to proceed along the monoclonal route.

Additionally impurities may be of very low abundance, but be highly antigenic in the host animal. In such a case, antigen-enriched preparations should be used for monoclonal antibody production. The screening method is sufficient to select out the antibodies of desired specificity.

2.3 Recombinant proteins as antigens

Often, the need for immunocytochemistry forms part of a wider project in plant molecular biology. In this case, cDNA clones corresponding to the antigen are available. A cDNA clone can express a protein in bacterial cells, either on its own or as a fusion protein attached to a carrier protein, such as beta-galactosidase or glutathione transferase (6). These proteins form effective antigens for preparation of polyclonal and monoclonal antibodies.

The greatest advantage of such systems is the abundance of the antigen, which may naturally be present at very low abundance in the native state in plant tissue. A second advantage is that the proteins are not glycosylated, so that only antibodies directed to peptide epitopes will be produced. A third advantage is that such proteins can be easily purified since clones can be engineered, for instance a flag peptide can be added to the cDNA sequence and immobilized antibodies against this flag peptide can be used for single-step purification of the antigen (7).

A further advantage is that heavy metal binding sites, such as a nickel-binding peptide sequence, can be engineered into the clone and the protein purified by means of a single run in a metal chelate column (8). Also, even if contaminating antibodies are present in such cases, they will be directed against bacterial components and not those of the native plant tissue. Polyclonal antibodies are quite suitable in such cases.

If a fusion protein is used to immunize an animal, the antibodies directed against the carrier protein can be removed by passing through an immuno-absorption column.

In certain cases, the recombinant protein is the only possibility for immunocytochemistry—especially in experiments where the antigen is obtained by expression of a cDNA clone obtained by subtractive hybridization. This is frequently the case when organ-specific transcripts are sought, when these are of unknown nature (9). Some cDNA clones do not express protein readily. In those cases, the deduced amino acid sequence can provide a source of synthetic peptides which can be prepared for immunization after conjugation to a suitable carrier molecules. Such antibodies raised to peptides, because they are directed against a specific epitope, may not be suitable for immunocytochemistry. These antibodies can be used for affinity purification of native antigens, which in turn can be used as immunogens for antibody production.

Examples of antibodies to fusion proteins and peptides which have been used successfully for plant antigen localization include the S- (self-incompatibility) specific glycoprotein of *Brassica* stigmas (10), the toxic proteins, thionins, of barley leaves that have anti-fungal properties (11), and a thiol protease, aleuraine, from barley endosperm (12).

2.4 Carbohydrates and glycoproteins as antigens

An unique feature of plant cells is their possession of rigid carbohydrate-rich cell walls. These are complex layered structures that should be ideal subjects for immunocytochemistry since their protein and carbohydrate antigens should be highly resistant to processing-induced localization artefacts (13). Apart from cell walls, carbohydrates also occur in dictyosome vesicles and starch granules. The latter are cytoplasmic components—produced within amyloplasts, a type of chloroplast.

Many protein antigens of plant origin are glycoproteins; they therefore contain glycan chains. A problem in the study of glycoprotein antigens is that similar glycan side chains may be present on a number of unrelated proteins. Polyclonal antisera raised against such a glycoprotein will not provide a specific immunocytochemical reagent because its use may result in artefactual labelling of different organelles and proteins. The antisera raised to glycoprotein antigens must be tested for specificity on Western blots of extracts of total cellular proteins. In Western blots, such antibody preparations will bind to a number of different proteins in the tissue extract. It may be possible to remove antibodies specific to carbohydrates by passing the serum through a carbohydrate affinity column (14), which will absorb out any cross-reactive antibodies. It will be more useful to deglycosylate the antigen preparation before immunization in the case of known glycoprotein antigens (15).

Monoclonal antibodies against an α-L-arabinoxyl residue of stylar glyco-proteins have been used to locate the antigen in the outer pollen tube walls (16). In plant extracts, arabinogalactan proteins are abundant molecules, are highly antigenic (17) and even in highly purified preparations of antigens, may be present as impurities. The presence of such components can often be seen as a background smear in Western blots. Again, it may be effective to absorb out such antibodies with immobilized arabinogalactans. Monoclonal antibodies offer the best alternative, as they can be selected for the antigen required. Antibodies can be selected which retain binding to deglycosylated antigens.

Recently, a protein component of starch granules from rye-grass pollen has been detected by immunocytochemistry. The antigen is a newly discovered allergenic protein (18). Monoclonal antibodies bind specifically to 31 kDa components within the starch granules (see *Figure 3*) suggesting that the function of this protein is involved with starch metabolism.

3. Standard protocol for plant cells

The standard techniques of immunolabelling cells both before and after embedding in resin are described here as they apply to most plant tissue types. Where special problems relating to embedding and sectioning or leaching of antigens from processed samples are encountered, the recommended methods will be found in Section 4.

3.1 Pre-embedding method

Immunolabelling of cells prior to their exposure to solvents and resins allows for the detection of macromolecules that may otherwise be denatured or masked during processing. The application of pre-embedding methods of plant studies have been limited. We present an outline of the current techniques used on plant tissue.

3.1.1 Whole cell and tissue labelling

Little work has been done on pre-embedding immunocytochemistry of whole plant cells due to the presence of cell walls that obstruct the penetration of gold probes. However, this approach may be most valuable for labelling of surface antigens, for example those of sperm cells which are natural protoplasts which have no detectable cell walls (19). Recently, isolated chloroplasts have been fixed and surface immunolabelled with an anti-idiotypic antibody prior to embedding and sectioning (20). This approach was used to identify an integral membrane protein as a receptor for protein import into the chloroplast stroma. Apart from isolated protoplasts, other plant materials with the potential for surface immunolabelling are pollen tubes, fungal spores and hyphae, all of which secrete substances into their surrounding milieu.

Immunolabelling of cytoskeletal components in plant cells and protoplasts, broken open in aqueous solution, has been performed without the need for post-embedding (21, 22). The thin layer of membrane—adherant cytoplasm remaining after cleaving is easily accessible to immunoprobes. Another technique which uses neither fixation nor embedding prior to immunolabelling has been applied to isolated membranes (23). Prepared by cell fractionation methods, the peribacterioid fraction from pea nodules was adhered to coated gold grids and labelled for bacterial and plant antigens prior to staining (see Chapter 4, *Protocols 14, 15*)

3.1.2 Pre-embedding labelling of sections

Thick cryosections of fixed, cryoprotected tissue, are thawed and labelled with primary antibody and immunogold reagents which enter the tissue by diffusion. The section is post-fixed, embedded, and sectioned for electron microscopy (24). This protocol gives good labelling of antigens near the surface of the tissue, but ultrastructure of underlying cells is poorly preserved and immunoprobes are restricted from penetrating membranes within the section. Consequently this protocol has not gained popularity as a standard technique. Another drawback is the movement of water-soluble substances which can occur during pre-treatment with a cryoprotectant and during immunolabelling.

3.2 Post-embedding method

The post-embedding technique involves fixing, dehydrating, and embedding the tissue in resin, followed by immunolabelling of the exposed antigen on the cut surface of sections. This method targets intracellular antigens and gives maximal resolution. It is by far the most commonly used technique for immunocytochemistry of plant cells and has recently been adapted for anhydrous preparations such as freeze-substitution.

3.2.1 Standard embedding for plant material

Fixation should preserve antigenicity, immobilize the antigen, and prevent its extraction, while at the same time ensure that satisfactory ultrastructure is maintained. The formulation of a fixative for immunolabelling is a compromise between good structure and good, accurate immunolabelling. For the great variety of plant structures, the precise fixation that best achieves this goal must be determined experimentally. A standard fixation regime may range from 0.5% glutaraldehyde for single cells, 2% paraformaldehyde and 1% glutaraldehyde for roots and stems, to 4% paraformaldehyde and 2% glutaraldehyde for some seeds.

The buffers most commonly used are Pipes (0.03–0.05 M) and phosphate (0.05 M) and occasionally sodium cacodylate (0.1 M), at pH 6.8–7.2. Fixation time may vary from 30 min at room temperature to overnight at 4 °C. Osmium fixation is not commonly used as it is incompatible with most resins used for immunolabelling. For instance osmium fixation prior to embedding in LR White resin has occasionally resulted in premature polymerization during infiltration. If osmium is used, it is etched off prior to immunolabelling by treatment with periodate/HCl (27, 28). Osmium may be used if polysaccharides are to be targeted.

Hydrophilic acrylic and methacrylate resins such as Lowicryl K4M, LR White, and LR Gold are most commonly used for immunocytochemistry. *Protocol 1* outlines the method used for embedding in LR Gold resin. Dehydration and infiltration times may need to be varied depending on the size of the sample and the extent of vacuolization of the plant tissue.

Protocol 1. Standard preparation of plant tissue for immunocytochemistry

1. Fix tissue as appropriate. During fixation, vacuum infiltrate for 30 min.

2. Wash twice in buffer for 30 min each.[a]

3. Wash in a series of decreasing concentrations of buffer to distilled water over 30 min.

4. Dehydrate through a graded series of ethanol in water 10%, 20%, 30%, 50%, 70%, 90%, 90%, for 20 min each change.

5. Infiltrate through a graded series of LR Gold in ethanol 10%, 20%, 50%, 70%, 90%, for 1 h each change.

6. Infiltrate in three changes of LR Gold resin for 3 h each change.

7. Place tissue in gelatin capsules, mounted in a metal block, and fill with a mixture of LR Gold resin and 1% benzyol peroxide paste (BPP).[b,c] After capping the gelatin capsule, the air bubble is removed by injecting more resin through the top with a fine syringe.

8. Allow a further 2 days before sectioning.

---

[a] All steps are best performed on a rotating wheel at room temperature to facilitate perfusion of the tissue.

[b] Concentrations of up to 3% BPP are required for hard-walled cells such as pollen, spores, and seeds.

[c] Fresh resin may polymerize too rapidly generating high temperatures which may denature antigens. A mixture of aged and fresh LR Gold resin may be needed to ensure that polymerization takes place slowly over a period of 24 h. Fresh LR Gold resin left at 4 °C in the original container gradually changes over a period of years such that the addition of the activator, benzoyl peroxide, results in slow or even no polymerization. The ratio of aged to fresh resin required for controlled polymerization needs to be determined experimentally.

3.2.2 Low temperature embedding

Embedding at room temperature may denature some antigens due to the heat produced upon polymerization of the resin. This can be reduced by polymerizing the resin at low temperature as outlined in *Protocol 2*. Alternatively, during dehydration in ethanol the temperature is progressively lowered to −30 °C prior to infiltration and embedding (29, 30). This slows the extraction of substances by the solvent, and leads to improved preservation of membranes.

Protocol 2. Embedding plant tissue at low temperature

1. After fixation, dehydration, and infiltration, add 0.5% benzil[a] to the LR Gold resin and infiltrate in the dark for 1 h.

2. Make up a fresh mix of LR Gold resin and benzil and fill gelatin capsules mounted in a standard 96-well microtitre tray. Insert the specimen and fit the lid. There is no need to exclude the air bubble.

3. Place the tray on aluminium foil at −30 °C about 15 cm below UV light. It is convenient to use the cutting chamber of a cryostat microtome for this purpose. Polymerize for 24 h.

4. Turn off the UV light and slowly bring to room temperature over 2 h. Allow a further 2 days before sectioning.

[a] Concentrations of up to 1% benzil are required for hard-walled cells such as pollen, spores, and seeds.

Some problems can be encountered using this technique. Large portions of tissue may not polymerize well at low temperature with air bubbles forming about the sample. This may be due to poor fixation and infiltration. However, low temperature embedding may improve the sectioning ability of smaller plant organs such as pollen and anthers.

The major problem encountered with sectioning is insufficient infiltration. This can be minimized by polymerizing only a few samples, testing their sectioning ability, and subsequently completing polymerization for all the samples or, if necessary, reintroducing ethanol into the liquid resin for further infiltration. '

3.2.3 Cryoultramicrotomy

More recently ultrathin cryosections have been prepared from fixed, cryoprotected tissue and labelled with antibodies and gold probes (see Chapter 4, Section 4). This is a specialized technique which requires the use of a cryoultramicrotome (25). Good immunolabelling and ultrastructural preservation have been achieved, but the problem with movement of water-soluble substances upon fixation, cyroprotection, thawing, and immuno-labelling persists. The application of this technique to botanical tissue has resulted in very poor sectioning properties (26).

3.2.4 Immunolabelling

Having immobilized the antigen and exposed it upon sectioning, the immunological reagents need to be applied under conditions that allow them to recognize the antigenic sites free of background and non-specific labelling. *Protocol 3* outlines the most commonly used method of immunolabelling.

High background labelling may be due to either a high concentration of antibody or ionic and/or electrostatic attraction (31). This can be resolved by one or more of several alternative treatments such as further diluting the antibody, adding bovine serum albumen throughout labelling, using Tween 20 in the primary blocking reaction, or by adding NaCl to the gold probe (see Chapter 10).

Protocol 3. Schedule for immunolabelling sections of specimens embedded in resin

1. Coat gold grids (200 mesh) with 1.8% Pioloform dissolved in filtered dry chloroform.

2. Cut sections, 80 nm thick, of embedded tissue and pick up on to grids and air-dry.

3. Float the grids on droplets of PBS (10 mM phosphate buffer with 150 mM NaCl, pH 7.2) containing 1% bovine serum albumin (BSA, Sigma Chemical Co., St Louis, USA), twice for 5 min each, to block non-specific binding of Ab.[a,b]

4. Incubate for 1 h in primary antibody diluted in PBS/BSA.

5. Wash on three drops of PBS for 5 min each. Wash on two drops of TBS (20 mM Tris–HCl, pH 8.2 with 150 mM NaCl) for 5 min each.

6. Incubate for 1 h on a goat IgG that recognizes the primary antibody, conjugated with 15 nm colloidal gold, diluted 1:15 with Tris-buffered saline.[c]

7. Wash on Tris-buffered saline twice for 5 min each.

8. Wash in 30 drops of distilled water from a squeeze bottle while holding the grid with jeweller's tweezers.

[a] All incubation steps are performed by floating grids, section-side down on droplets of reagent on Parafilm at room temperature. Humidity is maintained, if necessary, by covering the droplets with a petri dish lid and including a small container of water. Jeweller's tweezers should be dried with filter paper between washes to prevent grids from sinking into the reagents.
[b] 0.1% Tween 20 may be added for washing. If used, a further wash in PBS is required prior to incubation in antibody. Alternatively up to 0.5 M NaCl may be used.
[c] The antibody–gold complex can be centrifuged at 10 000g for 1 min to remove any aggregates of gold probes that may have formed during storage.

3.2.5 Silver enhancement

Silver enhancement (32) of immunolabelled sections enables the probe to be clearly seen on stained electron-opaque areas in the section. Low to medium power micrographs show clear labelling of the antigenic sites after silver enhancement. Gold particles are enhanced according to *Protocol 4* (see also

Chapter 4). Silver enhancement also enables the detection of multiple antigens on the same section (see Section 5, this chapter). If silver enhancement is to be performed for long periods, sections may need to be mounted on nickel grids.

Protocol 4. Silver enhancement of gold-labelled sections

1. Float grids on droplets of 0.5% gelatin (EIA grade reagent, Bio-Rad, California, USA) in water for 10 min.
2. Float grids on droplets of PBS/1% BSA for 10 min if necessary.
3. Wash grids in 30 drops of distilled water.
4. Blot off excess water.
5. Incubate on a drop of equal parts silver activator and silver initiator[a] for 20 min[b] to enlarge the particles to 60 nm.
6. Wash grids in 30 drops of distilled water and dry.
7. Stain on 2% aqueous uranyl acetate for 8 min in the dark.
8. Stain on aqueous Reynold's lead citrate for 1 min in a closed container containing NaOH pellets to absorb CO_2.

[a] Intense BL silver enhancement kit from Amersham (UK, USA).
[b] For sections with high labelling intensity, enhance for 10–15 min only.

3.2.6 Controls

Controls must be performed to determine the specificity of the immunolabelling reaction (33). Controls should be performed simultaneously with the standard labelling procedure. As many controls as possible should be performed to determine precise localization of the antigen.

i. Method control
This is performed by omitting the primary antibody step to test for the specificity of the gold probe.

ii. Antibody specificity controls
By mixing the primary antibody with purified antigen for 1 h prior to immunolabelling with this mixture, the level of receptor-mediated recognition of the antigenic sites on the section should be greatly reduced or absent. Purified antigen may not be available in some studies. The specificity of the experimental antibody can also be tested by immunolabelling sections with antibodies that are not specific for the tissue such as pre-immune or non-immune sera, or IgG of normal immunoglouulins at identical concentrations as the experimental antibody.

iii. Fixation controls

The use of both aqueous and anhydrous fixation is important for determining the amount of redistribution that the antigen may have undergone during aqueous fixation if the antigen is known to be water-soluble. For example, anhydrous fixation was necessary for the localization of rye-grass allergens in pollen (see Section 4.2).

iv. Tissue specificity controls

Another useful control is performed by immunolabelling tissue that is known not to have the antigen, such as mutant plant lines which do not express the antigen. Periodate treatment, which oxidizes sugar residues in sections, can be used as a negative control when labelling for sugars.

v. Double labelling

Multiple labelling of different antigens on the same section can give an indication of label specificity.

4. Special protocols for plant cells

The unique problems associated with handling plant tissue for immunocyto-chemistry can be overcome with the use of special protocols adapted for the particular types of tissue and antigen being investigated.

4.1 Highly vacuolate or fragile cells

When preparing highly vacuolate cells for immunolabelling, it is necessary to follow a fixation protocol that minimizes osmotic disruption to the cells. Vacuolate cells collapse or burst if they are fixed in vapour or under osmotic stress. A useful indicator of the conditions necessary for good fixation is gained by measuring the osmolarity of ground fresh tissue in a vapour-pressure osmometer. A measure of the relative osmolarity of cells can also be determined by watching for plasmolysis after placing the tissue in a decreasing series of sucrose in water. The relative osmotic values for fixatives can be determined from Hayat (34) for use on either delicate cells surrounded by their own plasma membrane, or for tissue with hard cell walls. Aldehyde fixation retains the semi-permeable properties of the plasma membrane. The cells become fully permeable only if osmium fixation is used.

Highly vacuolate or fragile tissue must be handled delicately to prevent collapse. For example, cells which show tip growth have a very fragile tip region which can easily burst when subjected to changes in osmolarity of the culture medium. When pollen tubes or fungal hyphae are grown *in vitro*, small changes in the osmolarity of the medium can cause the tube tips to burst and release their cellular contents. By fixing at a balanced osmolarity, and by avoiding osmotic shock during further processing, fragile cells such as pollen tubes can be successfully prepared for immunolabelling. *Protocol 5* is a schedule for embedding pollen tubes of *Brassica* for immunolabelling (*Figure 1*; reference 35).

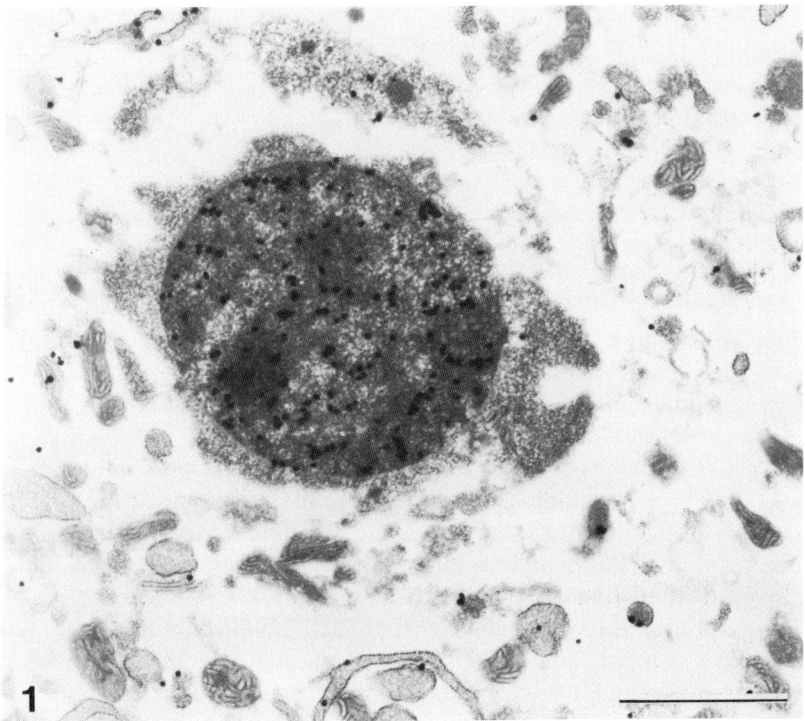

Figure 1. Immunogold localization of a *Brassica* sperm nuclear antigen in sperm isolated from germinating pollen tubes. Pollen tubes grown *in vitro* were osmotically ruptured and filtered. The sperm-enriched fraction was fixed in an osmotically balanced medium, embedded in LR Gold resin, and sections immunolabelled with a specific monoclonal antibody (MAb) (35). Primary MAb probed with 15 nm gold–goat anti-mouse IgG, enhanced with silver to 60 nm, and stained. Scale bar 1 μm.

Protocol 5. Fixation of pollen tubes of *Brassica napus* grown *in vitro*

1. Grow pollen tubes in a medium with 20% sucrose (35) and begin fixation with the addition of an equal volume of 2% glutaraldehyde in 0.05 M Pipes at pH 7.0, containing 16% sucrose. Fix for 20 min at room temperature.

2. Collect pollen tubes on to a 10 μm nylon filter.[a]

Protocol 5. *Continued*

3. Wash tubes in a decreasing series of sucrose in water 15%, 10%, 5% for 15 min each change.[a]

4. Dehydrate through a graded series of ethanol in water 10%, 20%, 30%, 40%, 50%, 60%, 70%, 15 min each change.[a] At this stage the tubes can be removed from the filter by opening the filter holder and gently agitating the filter in 70% ethanol.

5. Dehydrate in 80% and 90% ethanol for 20 min each change and infiltrate and embed according to *Protocol 1*.

---

[a] Pass all reagents through the filter with a syringe.

4.2 Desiccated cells

Plant tissue such as seeds and pollen, which must leave the parent plant upon maturity, are in a dormant and often dehydrated state. The low water content gives these tissues a high osmotic value and the component cells react rapidly when in contact with a moist environment. Pollen grains and seeds swell and pollen grains may even rupture when in contact with moisture. The localization of water-soluble proteins in these tissues can be achieved with the use of anhydrous fixation. To avoid the possible redistribution of water-soluble proteins, *Protocol 6* outlines the use of anhydrous fixation and processing of seed tissue. Anhydrous fixation at 4 °C is made possible with the use of 2,2-dimethoxypropane (DMP) (36). The DMP is able to precipitate water-soluble proteins and prevent their extraction. Water reacts with DMP in an equimolar equation (37) to produce acetone and methanol. After adding the fixatives, sufficient unreacted DMP must be left over to convert the water in the tissue into acetone and methanol. The DMP–fixative mixture is cooled to slow the reaction with the tissue. Other concentrations of glutaraldehyde and paraformaldehyde can be used to maintain ultrastructure, but the amount of water that can be introduced into the DMP is limited.

Protocol 6. The anhydrous preparation of *Sinapis* seed for the localization of myrosinase (38)

1. Cut whole mature seed into pieces less than 1 mm^3 in a mixture of 1% paraformaldehyde and 0.1% glutaraldehyde[a] in acidified DMP[b] at 4 °C and fix overnight.

2. Rinse in DMP, three changes for 30 min each at room temperature.

3. Rinse in a graded series of ethanol in DMP 30%, 50%, 70% for 30 min each.

4. Rinse in 100% ethanol three changes for 30 min each.

5. *En bloc* stain with 2% uranyl acetate in ethanol for 1 h.

6. Rinse in 100% ethanol three changes for 20 min each.

7. Rinse in a graded series of dimethylsulphoxide (DMSO) in ethanol 30%, 50%, 70%, 100% DMSO for 1 h each.

8. Infiltrate in a graded series of LR Gold resin in DMSO 20%, 40%, 60%, 70% for half a day each.

9. Infiltrate in 90% and 100% resin for 1 day each.

10. Infiltrate in three changes of resin for 1 day each and embed according to *Protocol 2*.

[a] Paraformaldehyde (Pf) is made as a 10% stock solution: 2 g of Pf in 20 ml water is heated to 60–65 °C and a few drops of 1 N NaOH is added until the solution clears and is adjusted to pH 7.2. Glutaraldehyde is obtained as 70% E.M. grade.
[b] 100 ml DMP with one drop of concentrated HCl.

The anhydrous processing of mature seed results in easy sectioning, good structural preservation, and precise immunolabelling (*Figure 2*). Silver enhancement of gold particles is not adversely affected by the *en bloc* staining with uranyl acetate because the uranyl acetate is probably extracted from the section surface during processing and immunolabelling. As the anhydrous processing of seed tissue utilizes low concentrations of fixatives, extraction in DMSO is necessary to make the tissue more permeable to the resin, thereby enabling better cross-linking upon polymerization and good-quality sections (39).

Mature *Sinapis* seed tissue was also fixed aqueously in 2.5% paraformaldehyde and 2% glutaraldehyde with 0.05 M Pipes buffer (38). The use of an *en bloc* stain with uranyl acetate followed by dehydration through a graded series of ethanol to 100% and a rinse in DMSO also improved the quality of sections from this tissue. The use of uranyl acetate and DMSO during processing did not affect immunolabelling intensity for the detection of myrosinase in seed tissue.

Tobacco seeds have been previously processed for immunolabelling from an aqueous fixative (30). After fixing in aldehydes and washing in buffer, the tissue was dehydrated in *N,N'*-dimethylformamide (DMF), and the tissue embedded in Lowicryl resin. DMF is a highly polar solvent which exracts pigment from the seed tissue and allows better polymerization of Lowicryl resin with UV light (30). Extraction of the antigens by DMF is not a problem as they are already locked into place by the cross-linking of the aldehyde fixatives. Replacing DMSO with DMF did not give very good sectioning quality for *Sinapis* seed tissue when fixed with either conventional aqueous or anhydrous conditions.

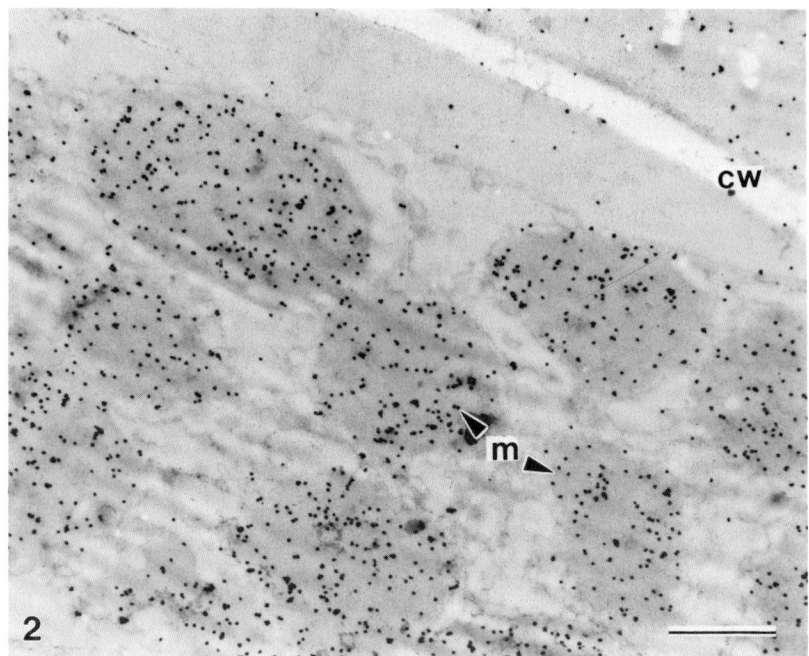

Figure 2. Immunogold localization of myrosinase in the cotyledons from a mature seed of *Sinapis*, processed by anhydrous fixation with glutaraldehyde, paraformaldehyde, and DMP. After *en bloc* staining with uranyl acetate, tissue was rinsed with DMSO and embedded in LR Gold resin (38). The sections were immunolabelled with rat polyclonal antibody to myrosinase, probed with 15 nm gold–goat anti-rat IgG, and silver enhanced to 40 nm. Myrosinase was located across the myrosin grain (m). Cell wall (cw). Scale bar 2 μm.

Mature seeds of tobacco have been sectioned and processed for the successful localization of storage proteins and a lectin following fixation in high concentrations of aldehydes (30, 40). This involved placing the seeds in 4% paraformaldehyde, 2% glutaraldehyde in 50 mM phosphate buffer at pH 7.0, removing the seed coat and dissecting out the radicle and cotyledons. High concentrations of fixatives were necessary to ensure good cross-linkage of the seed proteins. After dehydration in ethanol, the tissues were embedded in LR White resin at 65 °C for 2 days, which proved superior to Lowicryl embedding.

The hydrophobic embedding medium, Ladd LX-112 resin, may solve the problem of poor embedding of seeds while reducing the concentration of fixative required. Mature seeds of *Avena* have been aqueously fixed in 3% paraformaldehyde and 0.5% glutaraldehyde, embedded in LX-112 resin and successfully sectioned and probed for the localization of phytochrome (41). Compared with LR White resin, only LX-112 resin embedding was successful, probably due to the hydrophobic nature of the resin enabling penetration despite the high oil content of the embryos. Another hydrophobic resin, Lowicryl HM20, can be infiltrated and embedded at low temperature, and may also be useful for embedding plant tissues with a high oil content.

Phenolic compounds are common in tissue such as the over-wintering vegetative buds of deciduous trees. The difficulties of sectioning this tissue un-osmicated and embedded in LR white resin at 60 °C was overcome when polymerization was conducted at room temperature with a catalyst (P. E. Taylor, unpublished data). This is probably due to the dibutylphthalate in the catalyst stabilizing the phenolic compounds.

Mature pollen of ryegrass has been anhydrously processed for the detection of allergens that cause hayfever and asthma (36). The successful localization of the allergens depends upon anhydrous processing since the allergens of rye-grass pollen are water soluble and leach out quickly during hydration of the pollen grains in aqueous solution. *Protocol 7* outlines the use of glutaraldehyde vapour fixation and processing in DMP. Paraformaldehyde (Pf) could be used for vapour fixation, but this relies on using a resistor to heat the Pf in a closed container (42).

Protocol 7. Preparation of mature rye-grass pollen for the localization of water-soluble allergens

1. Fix whole, mature anthers with the vapour from 70% glutaraldehyde in a sealed container at room temperature for 24 h.

2. Dehydrate in acidified DMP[a] at 4 °C for 1 h.

3. Rinse in a graded series of ethanol in DMP 10%, 30%, 50%, 70%, 90%, 100% for 30 min each at room temperature.

4. Rinse in two changes of 100% ethanol for 30 min each.

5. Infiltrate in a graded series of LR Gold resin in ethanol 10%, 20%, 30%, 50%, 70%, 90%, 100% for 2 h each change.

6. Infiltrate in three changes of LR Gold resin for 3 h each change and embed according to *Protocol 2*.

[a] 100 ml of DMP with one drop of concentrated HCl.

This gives good ultrastructural preservation and adequate immunolabelling. An improvement in immunolabelling and sectioning was achieved, however, at the expense of structural integrity, by omitting vapour fixation and instead mixing the fixatives with DMP (*Figure 3*), as outlined in *Protocol 8* (18). Vapour fixation is not very successful with large or highly vacuolate tissue.

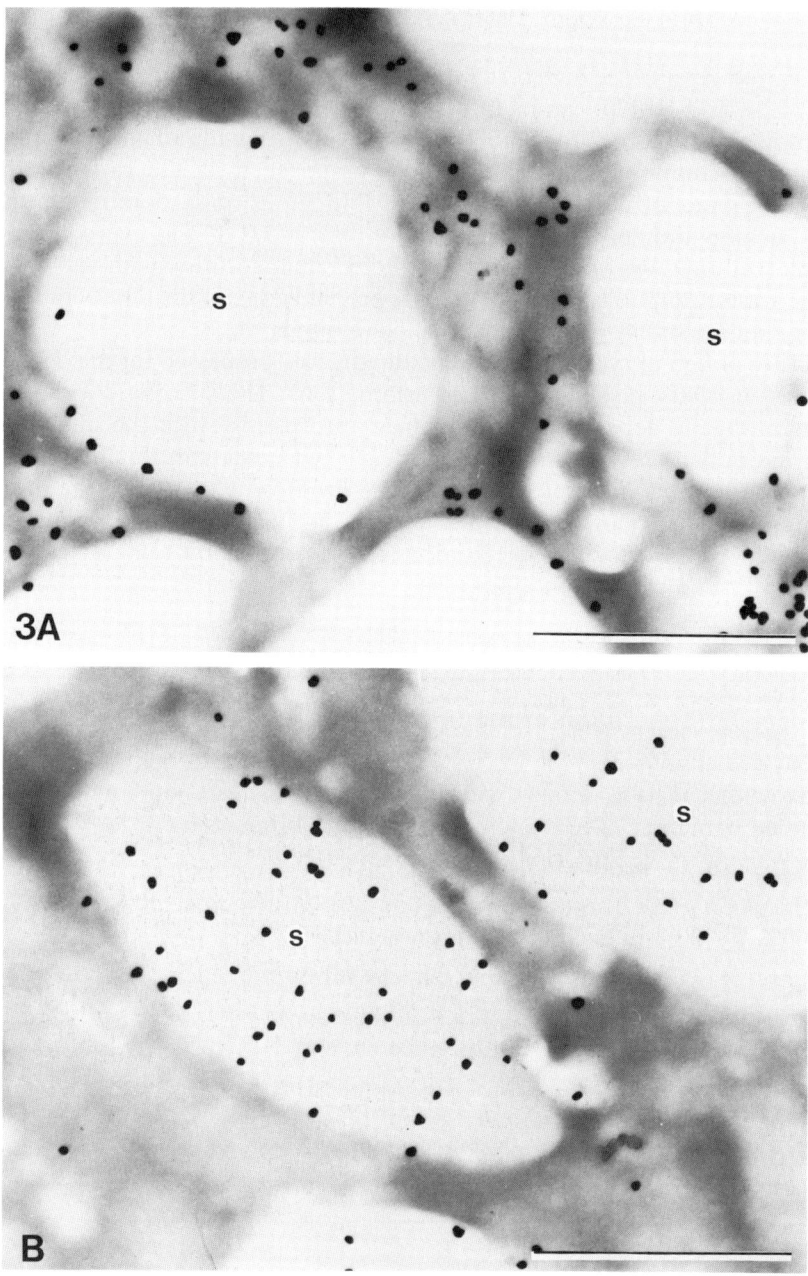

Figure 3. Immunogold localization of two different rye-grass pollen allergens, *Lol p*I and *Lol p*Ib. Pollen grains were processed by anhydrous fixation with glutaraldehyde, paraformaldehyde, and DMP and embedded in LR Gold resin. Sections were immuno-labelled with monoclonal antibodies (MAbs), 15 nm gold–goat anti-mouse IgG and silver enhanced to 60 nm (18). A, MAb specific for *Lol p*I labels the cytosol within the pollen grains. B, MAb specific for *Lol p*Ib labels mainly the starch granules (s). Scale bar 1 μm.

Protocol 8. An alternative fixation for the localization of allergens in rye-grass pollen

1. Fix whole anthers in a mixture of 1% paraformaldehyde and 0.1% glutaraldehyde in DMP at 4 °C for 4 h.
2. Rinse in DMP, three changes for 20 min each at room temperature.
3. Rinse in 100% ethanol and embed according to *Protocol 6*.

The DMP–fixative mixture has also been successfully used on the developmental stages of rye-grass pollen (P. E. Taylor, unpublished data). Fresh mature pollen can be easily collected into nylon bags (10 μm pore size. Swiss Screens, Australia) following anther dehiscence. Rye-grass pollen is only viable for 2 h in a dry atmosphere, or for up to 1 day in 100% humidity. The best immunolabelling and structural preservation may be achieved by combining vapour fixation with the DMP fix, or by the use of freeze-substitution.

Freeze-substitution has been used to investigate the localization of phytochrome in *Avena* coleoptiles (24), patatin in both potatoes (*Figure 4*; reference (14)) and transformed tobacco seed and leaf (26, 43). Freeze-substitution was necessary for the localization of S-protein in the papillar cells of *Brassica campestris* (44) and for immunolabelling actin in *Nicotiana* pollen

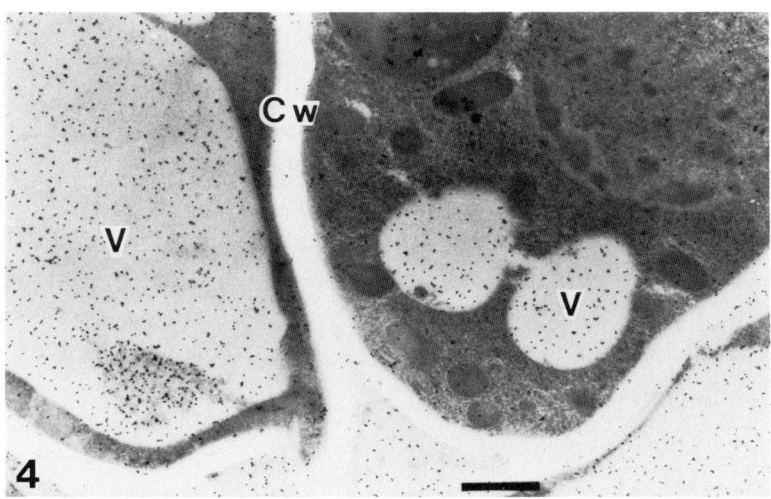

Figure 4. Immunogold localization of the storage protein patatin in potato tubers. Samples were prepared by high-pressure freeze-fixation, freeze-substitution in acetone containing uranyl acetate and glutaraldehyde and embedded in LR White resin. Sections were probed for patatin with a polyclonal antibody, followed by 15 nm gold–goat anti-rabbit IgG. The label was detected mainly in the vacuoles (V). Cell wall (Cw). Scale bar 1 μm. (From (14) by kind permission of Springer-Verlag.)

tubes (45). For freeze-substitution, tissue may be frozen against a copper block cooled in liquid helium, plunged in propane cooled by liquid nitrogen, or frozen under high pressure in liquid nitrogen (46). Substitution-fixation is generally performed at −80 °C in acetone and osmium, or acetone, uranyl acetate and glutaraldehyde. After a rinse in 100% ethanol at room temperature, the samples are embedded in LR White or epoxy resins. Prior to immunolabelling, osmicated sections are etched in H_2O_2 or sodium metaperiodate–HCl (27, 28). This maintains ultrastructure and allows immunolabelling of some water-soluble proteins.

Without the use of osmium, plant tissue tends to be poorly infiltrated after freeze-substitution and is consequently difficult to section. We have recently developed a protocol for freeze-substitution of pollen for immunolabelling. Pollen is well suited to freezing due to its low water content and small size.

Outlined in *Protocol 9* is a simple freeze-substitution method developed for the localization of allergens in rye-grass pollen adapted from Kaeser (37). The use of DMP for freeze-substitution eliminates the need to change solutions at low temperature for removal of tissue-bound water. The use of uranium was necessary to enable sectioning. (See also Chapter 4 for further details of freezing, freeze-substitution, and post-embedding immunolabelling.)

Protocol 9. Simple freeze-substitution method for the localization of water-soluble allergens in mature rye-grass pollen

1. Load mature pollen into copper sandwiches[a] and plunge-freeze in liquid propane cooled in liquid nitrogen.

2. Place samples in a centrifuge tube with a mixture of 0.5% glutaraldehyde, 0.5% paraformaldehyde, 0.5% uranyl acetate, 25% acetone, 20% methanol, and 50% acidified DMP[b] frozen in liquid nitrogen.

3. Store at −70 °C in a freezer for 1 day.[c]

4. Move to −30 °C for 1 day in a chest-freezer.

5. Move to −10 °C for 4 h in a fridge-freezer.

6. Move to 4 °C for 1 h in a refrigerator.

7. Move to room temperature for 1 h.

8. Rinse in DMP for 15 min.

9. Rinse in 100% ethanol and embed according to *Protocol 7*.

[a] Specimen carriers are from Balzers (Germany, UK), and have been made from copper sheet hollowed to 50 μm thickness (46).

[b] Glutaraldehyde is obtained as 70% EM grade, paraformaldehyde is from 10% aqueous stock solution, uranyl acetate is from 10% uranyl acetate in methanol, acetone and methanol are dried on molecular sieve, and DMP is made fresh on the day.

[c] At −70 °C the fixative mixture melts, the copper sandwich sinks into it, and the substitution-fixation takes place with the conversion of water in the sample into acetone and methanol.

The freeze-substitution method given in *Protocol 9* combines good structure and immunolabelling on the same material, for example, in rye-grass pollen (*Figure 5*). After sectioning, the grids are floated on water for 30 min to remove hydrostatically bound uranyl acetate from the section prior to immunolabelling. However, pollen grains that were most fixed by the uranium and thus gave superior sections were much reduced in immunolabelling intensity. A rinse in DMSO may be utilized to improve infiltration of pollen, such as *Brassica*, that is rich in lipids.

To localize water-soluble phytohormones such as abscisic acid (47) and indole-3-acetic acid (48), reagents which can cross-link low molecular weight plant hormones with cellular proteins have been employed on tissues prior to fixation and embedding in Lowicryl resin. Various plant tissue types are immersed in a 2% aqueous solution of water-soluble 1-(3-dimethylamino-propyl)-3 ethylcarbodiimide which reacts with the carboxyl group of the hormone to immobilize it. The tissues are embedded and the hormone localized with standard immunocytochemical methods. Aqueous precipitation

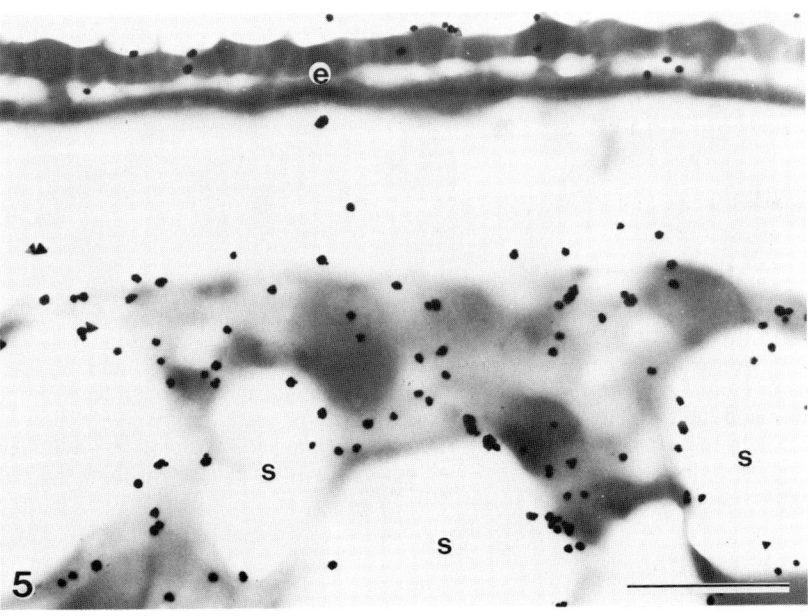

Figure 5. Immunogold localization of rye-grass pollen allergenic protein *Lol p*I in freeze-substituted grains. Samples were plunge-frozen in liquid propane, and freeze-substituted in acetone containing DMP, methanol, uranyl acetate, glutaraldehyde, and para-formaldehyde before embedding in LR Gold resin. Sections were immunolabelled with monoclonal antibodies specific for *Lol p*I and 15 nm gold–goat anti-mouse IgG followed by silver enhancement to 60 nm. Immunolabel was detected in the cytosol and exine (e) but not in the starch granules (s). (Unpublished data, P. E. Taylor.) Sections were not post-stained. Scale bar 1 μm.

techniques for water-soluble proteins have had limited success, especially when applied to pollen grains due to the rehydration-induced structural changes and subsequent movement of antigens prior to the immobilization by fixation. Recently, brassinosteroids have been localized in *Brassica* pollen tubes when processed according to *Protocol 5* (49). Although fixed aqueously, germinated pollen showed storage of this phytohormone in plastids (*Figure 6*). Movement of the antigens upon fixation may have been limited as the pollen tubes were grown *in vitro* in an aqueous medium.

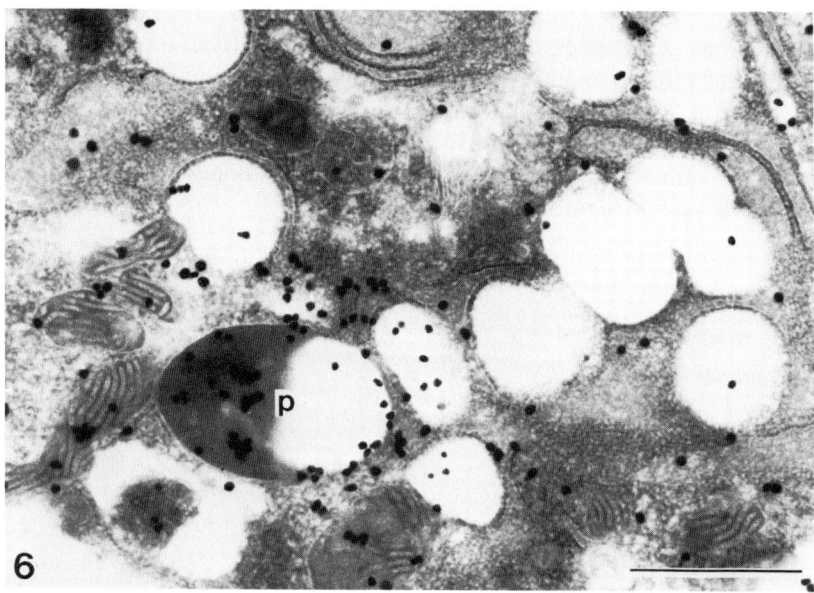

Figure 6. Immunogold localization of brassinosteroids in the pollen tubes of *Brassica napus*, grown *in vitro*. Germinated pollen was fixed aqueously in an osmotically buffered medium and embedded in LR Gold resin. Sections were probed with polyclonal antibodies to castasterone followed by 15 nm gold–goat anti-rabbit IgG, silver enhanced to 60 nm and stained. Immunolabel was detected within plastids (p) and the nearby cytosol (49). Scale bar 1 μm.

4.3 Organelles

Osmium fixation is not generally recommended for tissues prior to embedding in acrylic or methacrylate resins. On un-osmicated tissue, membrane preservation is poorly visualized after sectioning, particularily in the plastid envelope, Golgi apparatus, plasma membrane, and protein vacuoles. This may be overcome by *en bloc* staining in uranyl acetate after the primary fixation but prior to dehydration. If residual uranium prevents silver enhancement and to enable membrane associated antigens to be localized, sections should be washed in water prior to immunolabelling to remove hydrostatically bound uranium. Fixation with Pipes buffer and low temperature embedding have been found to be the most successful methods for the preservation of organelle membranes (13, 50). Isolated organelles and protoplasts should be processed in an osmotically balanced buffer according to *Protocol 5* (*Figure 1*). The handling of plant fractions or small cells can be performed by embedding in low temperature agarose (29), by pelleting, by collecting on to nuclepore filters with 1 μm pore size (Nuclepore, Pleasanton, California) or in bags made from nylon screen.

4.4 Cell walls

Cell walls contribute to difficulties experienced in infiltration and sectioning. For secondarily thickened cell walls, the embedding resin should be adjusted to give a hardened consistency. The length of fixation and infiltration needs to be extended to meet the requirements of hard-walled cells. For example, bark has been most successfully embedded for immunolabelling in the hard grade of LR White resin (30).

4.5 Cuticles, surface waxes, and exudates

Leaves are often covered with a thick layer of cuticle and waxes, the stigma with an exudate, and pollen with a lipidic pollencoat. These often result in sections pulling away at the surface/resin interface. The problem may be overcome by the use of phase partition fixation in which the fixative is shaken with heptane and the anhydrous phase is used initially to fix plant material as shown in *Protocol 10* (51). This has been compared to vapour fixation as the heptane penetrates the cuticle and cell wall but stops at the plasma membrane, whereby the fixative alone penetrates further into the cytoplasm. Disruption of the cell's ionic balance may be minimal. Phase partition fixation gives good structural preservation and may improve infiltration. Vacuoles are notoriously difficult to fix but this method may satisfactorily preserve large central vacuoles. Embedding in the hydrophobic resin Lowicryl HM20 may help solve the problems of processing plant material coated in hydrophobic substances. Fixation with uranyl acetate and embedding in Lowicryl HM20 resulted in sections of high quality (52).

Protocol 10. Phase partition fixation of plant material

1. Mix freshly prepared 4% paraformaldehyde and 0.25% glutaraldehyde in 0.05 M Pipes buffer, pH 7.3. Shake 50 ml of fixative with an equal volume of *n*-heptane for 1 min and let stand until the phases separate. Carefully decant the heptane (upper) phase.

2. Dissect plant material in the heptane phase.

3. Remove samples from the heptane and complete fixation and processing according to *Protocol 1*.

---

5. Localization of multiple antigens in the same tissue section

Double immunogold labelling of the same section with two different antibodies of the same isotype has been performed to detect different allergenic proteins in rye-grass pollen (*Figure 7*; reference (18)). After the

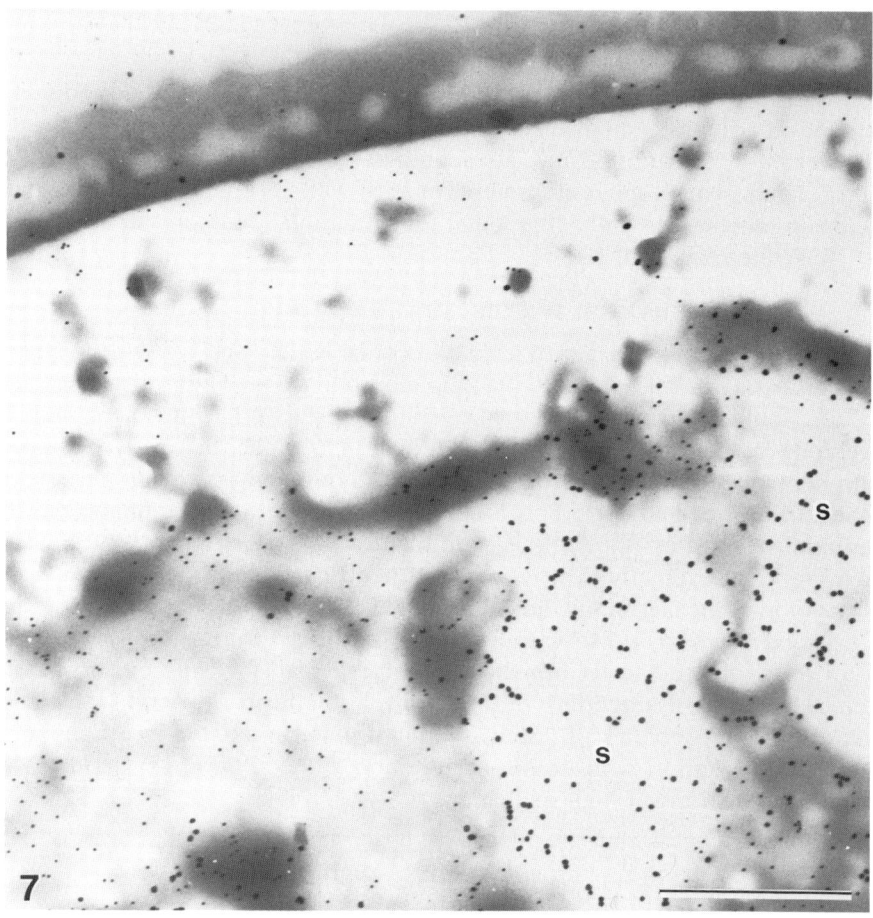

Figure 7. Immunogold localization of rye-grass pollen allergenic proteins *Lol p*I and *Lol p*Ib by double labelling of sections with same sized gold probes but with the first gold probe silver enhanced. Tissue processed as described in *Figure 3*. Sections were first incubated in monoclonal antibodies (MAbs) specific for *Lol p*Ib and 15 nm gold–goat anti-mouse IgG, followed by silver enhancement to 60 nm. The same section was then incubated in MAbs specific for *Lol p*I and 15 nm gold–goat anti-mouse IgG. The large particles show the cellular sites of *Lol p*Ib—mainly in the starch granules (s), while the small particles show location of *Lol p*Ib in the cytosol. No post-staining to enhance electron contrast (18). Scale bar 1 μm.

first immunolabel is applied, the gold probe is silver enhanced to mask specific antigenic sites on the antibody–gold complex. The second label and gold probe are applied to target other specific antigenic sites on the section. Enhancement is not performed after application of the second probe. This results in two different sized markers on the section surface (*Figure 2*). On a separate section, the order of the antibodies is reversed to check for the specificity of the immunolabelling.

The localization of secreted adhesive material on *Phytophthora* zoospores (29) has been explored using double immunolabelling. Here the immuno-labelling is direct using antibodies complexed to the gold probe (*Figure 8*). Multiple labelling is performed on the same section as marker sizes are different. Antibodies of the same isotype are used since the antibody–gold complexes do not interact with each other.

For a detailed account of multiple immunolabelling techniques refer to Chapter 6.

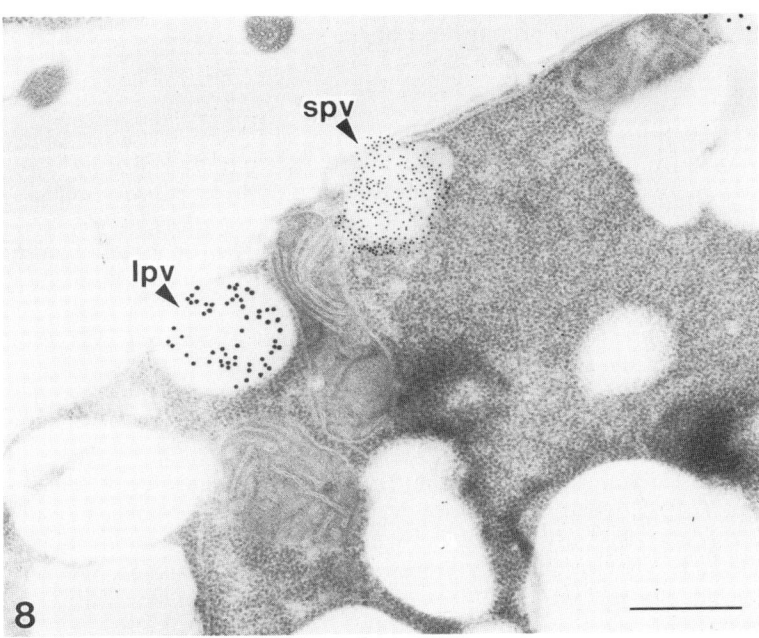

Figure 8. Immunogold localization of adhesive components secreted from zoospores of the protist *Phytophthora* during encystment by double labelling of sections with two different sized antibody–gold probes. The zoospores were embedded in low temperature agarose prior to standard aldehyde fixation and low temperature embedding in Lowicryl K4M. Two different monoclonal antibodies bind to components of large peripheral vacuoles (lpv) and small peripheral vasicles (spv). Scale bar 0.5 µm (From (29) by kind permission of The Company of Biologists Limited).

6. Localization of antigens in transgenic plants

An exciting area of immunocytochemistry is its potential applications in understanding the expression of introduced genes in transgenic plants; to date, exclusively in transgenic tobacco plants. This technique can provide otherwise unobtainable information on the sub-cellular sites of protein accumulation, and may be important in understanding the expression of developmentally regulated genes. Examples include the localization in transgenic tobacco tissues of zein from maize, in protein bodies (40), vacuolar phytohaemaglutinin from French bean (53), the potato storage protein patatin (43), and the extracellular S-proteins of *Brassica* (44, 54).

Acknowledgements

We wish to thank the Australian National Health and Medical Research Council, the Australian Research Council for financial support, and Terryn Hough for assistance in preparation of the manuscript.

References

1. Knox, R. B. (1982). In *Immuncytochemistry*, Vol. 1 (ed. G. Bullock and P. Petrusz), p. 205. Academic Press, London.
2. Knox, R. B. and Clarke, A. E. (1978). In *Electron Microscopy and Cytochemistry of Plant Cells*, (ed. J. L. Hall), p. 150. Elesvier/North Holland, Biomedical Press, Amsterdam.
3. Knox, R. B. and Singh, M. B. (1985). In *Botanical Microscopy 1985*, (ed. A. W. Robards), p. 208. Oxford University Press.
4. Hawes, C. (1988). In *Plant Molecular Biology. A Practical Approach*, (ed. C. H. Shaw), p. 103. IRL Press, Oxford.
5. Smith, E. D. and Fisher, P. A. (1984). *J. Cell Biol.*, **99**, 20.
6. Knapp, S. Broker, M., and Amann, E. (1990). *Biotechnology*, **8**, 280.
7. Hopp, T. P., Prickett, K. S., Price, V. L., Libby, R. T., March, C. J., Cerretti, D. P., *et al.* (1988). *Biotechnology*, **6**, 1204.
8. Hochuli, E., Bannworth, W., Dobeli, H., Gentz, R., and Stuber, D. (1988). *Biotechnology*, **6**, 1321.
9. Brown, S. M. and Crouch, M. L. (1990). *The Plant Cell*, **2**, 263.
10. Umbach, A. L., Lalonde, B. A., Kandasamy, M. K., Nasrallah, J. B., and Nasrallah, M. E. (1990). *Plant Physiol.*, **93**, 739.
11. Bohlmann, H., Clausen, S., Behnke, S., Giese, H., Hiller, C., Reimann-Philipp, U., *et al.* (1988). *EMBO J.*, **7**, 1559.
12. Holwerda, B. C., Galvin, N. J., Baranski, T. J., and Rogers, J. C. (1990). *The Plant Cell*, **2**, 1091.
13. Herman, E. M. (1988). *Ann. Rev. Plant Physiol. Mol. Biol.*, **39**, 139.

14. Sonnewald, U., Studer, D., Rocha-Sosa, M., and Willmitzer, L. (1989). *Planta*, **178**, 176.
15. Edge, A. S. B., Faltynek, C. R., Hof, L., Reichert, L. E., and Weber, P. (1981). *Anal. Biochem.*, **118**, 131.
16. Clarke, A. E., Anderson, M. A., Bacic, T., Harris, P. J., and Mau, S. L. (1985). *J. Cell Sci. Suppl.*, **2**, 261.
17. Anderson, M. A., Sandrin, M. S., and Clarke, A. E. (1984). *Plant Physiol.*, **75**, 1013.
18. Singh, M. B., Hough, T., Theerakulpisut, P., Avjioglu, A., Davies, S., Smith, P. et al. (1991). *Proc. Natl. Acad. Sci. USA*, **88**, 1384.
19. Taylor, P. E., Kenrick, J., Blomstedt, C. K., and Knox, R. B. (1991). *Sex. Plant Reprod*, **4**, 226.
20. Pain, D., Kanwar, Y. S., and Blobel, G. (1988). *Nature*, **331**, 232.
21. Traas, J. A. and Kengen, H. M. P. (1986). *J. Histochem. Cytochem.*, **34**, 1501.
22. Goodbody, K. C., Hargreaves, A. J., and Lloyd, C. W. (1989). *J. Cell Sci.*, **93**, 427.
23. Bradley, N. J., Butcher, G. W., Galfre, G., Wood, E. A., and Brewin, N. J. (1986). *J. Cell Sci.*, **85**, 47.
24. McCurdy, D. W. and Pratt, L. H. (1986). *J. Cell Biol.*, **103**, 2541.
25. Tokuyasu, K. T. (1986). *J. Microsc.*, **143**, 139.
26. Greenwood, J. S. and Chrispeels, M. J. (1985). *Plant Physiol.*, **79**, 65.
27. Craig, S. and Goodchild, D. J. (1984). *Protoplasma*, **122**, 35.
28. Vandenbosch, K. A. and Newcomb, E. H. (1986). *Planta*, **167**, 425.
29. Gubler, F. and Hardham, A. R. (1988). *J. Cell Sci.*, **90**, 225.
30. Herman, E. M. (1989). In *Colloidal Gold. Principles, Methods and Applications*, Vol. 2, (ed. M. Hayat), p. 304. Academic Press, San Diego.
31. Boenisch, T. (1990). In *Immunochemical Staining Methods*, (ed. S. Naish), p. 21. DAKO Corp., California.
32. Danscher, G. and Norgaard, J. O. R. (1983). *J. Histochem. Cytochem.*, **31**, 1394.
33. Pool,, C. W., Buijis, R. M., Swaab, D. F., Boer, G. J. and Leeuwen, F. W. van (1983). In *Immunocytochemistry*, (ed. C. A. Cuello), p. 1. Wiley, Chichester.
34. Hayat, M. (1981). *Fixation for Electron Microscopy*. Academic Press, New York.
35. Kenrick, J., Blomstedt, C. K., Taylor, P. E., Singh, M. B., and Knox, R. B. (1992). *Protoplasma* (In press.)
36. Staff, I. A., Taylor, P. E., Smith, P., Singh, M. B., and Knox, R. B. (1990). *Histochem. J.*, **22**, 276.
37. Kaeser, W. (1989). *J. Microsc.*, **154**, 273.
38. Dungey, S. and Taylor, P. E. in preparation.
39. Kim, E. S. and Mahlberg, P. G. (1991). *Am. J. Bot.*, **78**, 220.
40. Herman, E. M., Chrispeels, M. J., and Hoffman, L. M. (1989). *Cell Biol. Int. Rep.*, **13**, 37.
41. Leurentop, L. and Verbelen, J-P. (1989). *Micron. Microsc. Acta*, **20**, 131.
42. Newman, T. and Briarty, L. G. (1990). *Ann. Bot.*, **65**, 305.
43. Sonnewald, U., Sturm, A., Chrispeels, M. J., and Willmitzer, L. (1989). *Planta*, **179**, 171.
44. Kishi-Nishizawa, N., Isogai, A., Watanabe, M., Hinata, K., Yamakawa, S., Shojiam, S. et al. (1990). *Plant Cell Physiol.*, **31**, 1207.
45. Lancelle, S. A. and Hepler, P. K. (1989). *Protoplasma*, **150**, 72.

46. Gilkey, J. C. and Staehelin, A. (1986). *J. Electron Microsc. Tech.*, **3**, 177.
47. Sossoutzov, L., Sotta, B., Maldiney, R., Sabbagh, I., and Miginiac, E. (1986). *Planta*, **168**, 471.
48. Ohmiya, A., Hayashi, T., and Kakiuchi, N. (1990). *Plant Cell Physiol.*, **31**, 711.
49. Sasse, J., Yokota, T., Taylor, P. E., Griffiths, P., Porter, Q., and Cameron, D. (1991). In *Progress in Plant Growth Regulation.* (ed. C. M. Karssen, L. C. Van Loon, and D. Vreugdenhil), p. 319. Kluwer, Netherlands.
50. Staff, I. A., Taylor, P. E., Kenrick, J., and Knox, R. B. (1989). *Sex. Plant Reprod.*, **2**, 70.
51. McFadden, G. I., Bonig, I., Cornish, E. C., and Clarke, A. E. (1988). *Histochem. J.*, **20**, 575.
52. Benichou, J. C., Frehel, C., and Ryter, A. (1990). *J. Electron. Microsc. Tech.*, **14**, 289.
53. Sturm, A., Voelker, T. A., Herman, E. M., and Chrispeels, M. J. (1988). *Planta*, **175**, 170.
54. Kandasamy, M. K., Dwyer, K. G., Paolillo, D. J., Doney, R. C., Nasrallah, J. B., and Nasrallah, M. E. (1990) *The Plant Cell*, **2**, 39.

6

Multiple immunolabelling techniques

JULIAN E. BEESLEY

1. Introduction

Comparison of several cellular components in one specimen is an exciting possibility introduced by the multiple immunolabelling technique. The challenge of multiple immunolabelling therefore is to immunolabel two or more antigenic sites, each with a distinct probe, so that the site of each antigen of interest can be unequivocally determined in a single preparation. Theoretically, any antigen that can be immunolabelled either by light or electron microscopy can be detected in multiple immunolabelling studies. Any number of antigens may be immunolabelled in one specimen provided distinct probes are used to localize each antigen. Probes for multiple immunolabelling should provide sufficient contrast to be easily distinguishable from each other at all magnifications. Multiple immunolabelling techniques are combinations of the single immunolabelling techniques given elsewhere in this book (Chapters 3 and 4). There are many different approaches to achieving successful multiple labelling, both in terms of the probe used to localize the antigens of interest and the nature and sequence of antibodies applied. The most common methods for multiple immunolabelling are given. Potentially any combination of different probes could be used, and since multiple immunolabelling depends on the availability of suitable reagents any combinations of immunolabelling should be considered. Although the philosophy of multiple immunolabelling is identical for both light and electron microscopy and many similarities exist between the techniques, there are distinct technical differences and therefore light and electron microscope techniques are described separately. The methods given in this chapter describe the immunolabelling techniques rather than specimen preparation techniques and the reader is referred to Chapters 3 and 4 for details of specimen preparation to obtain a specimen for immunolabelling.

2. Selection of the technique

2.1 Light or electron microscopy

For multiple immunolabelling the nature of the problem and the required resolution will have been defined by single immunolabelling (Chapters 3 and 4) and either light or electron microscope multiple immunolabelling should be approached directly.

2.2 Species of primary antibody

The major consideration when choosing the immunological reagents for multiple immunolabelling is the possibility of cross reactions between reagents. There is usually little choice in the primary antibodies for multiple immunolabelling studies. It is those which are available and which have been successfully used for single immunolabelling. A multiple immunolabelling technique should be chosen which minimizes cross reactions between reagents. This does not necessarily imply that only antibodies from different host species can be used or that immunological reagents with cross reactions cannot be used. Techniques exist for multiple immunolabelling with antibodies from the same or different host species and for immunolabelling with probes directed at the same Ig class. Any antibody, whether monoclonal or polyclonal, can be used for multiple immunolabelling.

2.3 Position of antigenic sites

The relative proximity of the different antigens to be localized must be considered. Immunolabelling different antigens which are present on different cells or structures is usually not problematical. Difficulties arise when the antigens of interest occur on the same structure or cell. If two antigens are very close on the specimen the accumulation of reagents used for immuno-labelling the first antigen may prevent immunolabelling of the second antigen by physically blocking access of the reagents to the second antigen.

2.4 Immunolabelling techniques

The schedules employed for multiple immunolabelling are identical to those already described for single immunolabelling studies (Chapters 3 and 4). If antigen–antibody reactions of differing sensitivities are being used, it is imperative that the specimen is prepared to accommodate the most sensitive antigen.

Multiple immunolabelling schedules are either sequential if cross reactions between reagents are suspected or simultaneous if cross reactions do not exist. The use of primary antibodies raised in different species, a situation which is common with polyclonal antibodies, facilitates simultaneous multiple labelling since problems associated with cross reactivity between the immunolabelling reagents are avoided. Immunolabelling can therefore be performed by incubating the specimen with mixtures of both antibodies followed by mixtures of both microscopically dense probes. The technique takes no longer than a single immunolabelling schedule.

The majority of monoclonal antibodies are raised in mice and therefore there are frequent occasions when two or more primary antibodies of the same species have to be used for multiple immunolabelling. Sequential application of reagents in combination with adequate blocking measures is necessary to minimize cross reactions.

The possibility of steric hindrance should always be considered and it is always necessary to perform immunolabelling with probes optimally diluted as judged by single immunolabelling experiments, so that the minimum number of antibodies are present. The optimal reagent concentrations will remain identical when multiple immunolabelling. If simultaneous immunolabelling is performed the reagents applied in each step must be optimized to the same time interval. Care must also be taken to prepare mixtures of reagents with the correct final concentration.

2.5 Controls

It is assumed in this chapter than adequate controls will have been performed during single immunolabelling of each antigen to test the performance and specificity of each of the reagents. There is a high risk of cross contamination in all multiple immunolabelling schedules even if totally different species of reagents are used. It is most important therefore to perform the following controls to achieve confidence in the results.

(a) Omission of primary antibodies from successive layers and their substitution with non-immune serum or inappropriate antibodies.

(b) Reversal of the order of antigen detection to ensure that masking of one antigen by immunolabelling another is not occurring. Antigens which are present in low amounts should be visualized prior to the more abundant antigen to ensure masking does not occur.

(c) Inclusion of a known positive control specimen for all antigens of interest.

3. Light microscopy

3.1 Choice of probes

The majority of images for light microscope immunocytochemistry are obtained through coloration of the antigenic sites. Resolution of the probes is not usually a problem for light microscopy which is used to define immunolabelling of whole cells and at best to designate antigens to parts of cells. For light microscopy the choice of coloured probes is wide. Fluorochromes and enzymes are routinely employed and their use depends upon the availability of a fluorescent microscope, the relative positions and abundance of the antigenic sites, and the requirement for long-term storage of slides. The epi-illumination and dark field techniques are occasionally used to image colloidal metal markers to contrast with the coloured reaction products of the enzyme systems. *Figures 1–3* are presented as typical examples to demonstrate the potential of the technique.

The fluorochromes were the first probes successfully utilized in immuno-cytochemistry (1). Since the early development of fluorescein isothiocyanate (FITC) (2) as an apple-green fluorescing probe a number of other compounds have been introduced which, on excitation by ultraviolet light, emit at different wavelengths to FITC and produce contrasting colours such as red (tetramethylrhodamine isothiocyanate (3), lissamine sulphonyl chloride (4), or Texas red (5)), blue (diethylaminocoumarin (6)) or yellow orange (phycoerythrin (7)) and have been successfully used for multiple immuno-labelling. If antigens are present in greatly differing quantities the fluorescent probes are particularly useful since each probe can be visualized separately without hindrance from the other by use of different wavelengths of light for excitation and detection.

A range of contrasting colours can be produced using enzyme probes by development of the enzyme with different chromogens. Peroxidase can be developed to brown with 3'3 diaminobenzidene (8), to red with 3-amino-9-ethylcarbazole (9), or to dark blue-purple with 4-chloro-1-naphthol (10) and alkaline phosphatase can be developed to red with Fast Red TR (11), or New Fuchsin (12) or to blue with Fast Blue BB (11) and to violet with Fast Red Violet LB (9). Beta-galactosidase is developed to blue-green with potassium ferro/ferricyanide (13) and glucose oxidase to blue-purple with nitroblue tetrazolium (14). Many combinations of these enzyme labels have been used for multiple immunolabelling. The reaction product from one enzyme probe can mask that of another, particularly if the antigens are present in greatly differing quantities. They are most successful for detecting antigens on different cells or structures. Some investigators, however, have reported successful visualization of two different antigens on the same cell in which the combination of two reaction products produces a third colour which is distinguishable from each individual chromogen. Interpretation of these colour combinations can be problematical particularly if one antigen is present in excess over the other.

The silver enhancement technique performed with colloidal gold probes is a useful marker for use in conjunction with an enzyme technique (15), the silver enhancement technique forming a black reaction product proving expecially useful for the visualization of sparse antigens. The technique may also be used in conjunction with colloidal gold probes alone, the red of the colloidal gold contrasting with the black of the silver enhanced probes (16). In addition Roth 1982 (17) demonstrated the use of red colloidal gold and yellow colloidal silver probes for multiple immunolabelling. Although

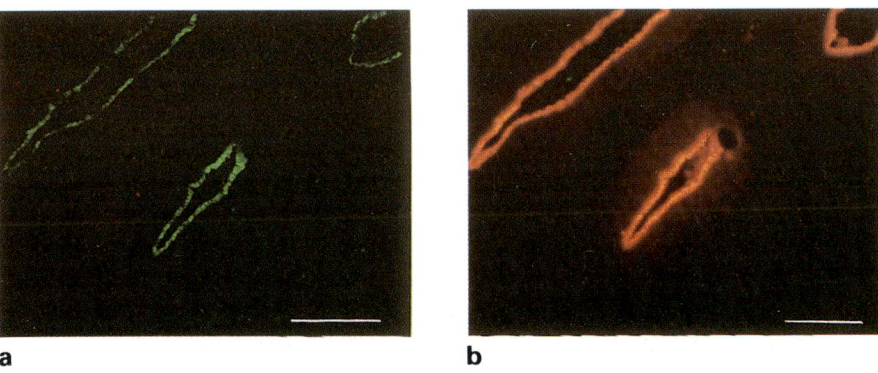

a   b

Figure 1. Cryostat section of fibrocystic disease from breast tissue was incubated sequentially with a mixture of antibodies to Ca1 (IgM) and to HMFG2 (IgG) followed by a mixture of FITC-labelled anti-mouse IgM and TRITC-labelled anti-mouse IgG as described in *Protocol 6*. Examination with a fluorescent microscope equipped with an excitation filter BP 450–490 and barrier filter BP 520–560 for FITC and an excitation filter BP 546/12 and barrier filter LP 590 for TRITC indicated that immunolabelling for both Ca1 (a) and HMFG2 (b) occurred in similar groups of cells. (Kindly reproduced from Beckford, U. *et al.*, (1985). *J. Clin. Pathol.*, **38**, 512–20.) Scale bar 800 μm.

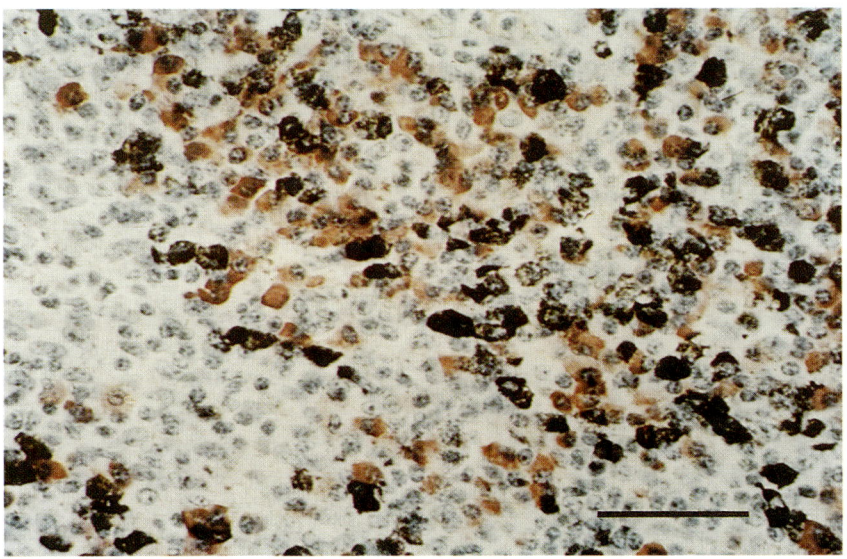

Figure 2. A section of formalin-fixed and paraffin-embedded tonsil. Plasma cells demonstrating IgG (black) were localized with the silver enhancement method and plasma cells demonstrating IgA (red) were localized with the indirect peroxidase method using 3-amino-9-ethylcarbazole as the chromogen as outlined in *Protocol 3*. (Kindly supplied by Mr P. Jackson, Academic Unit of Pathology, Department of Clinical Medicine, School of Medicine, The University of Leeds, Leeds, UK.). Scale bar 50 μm.

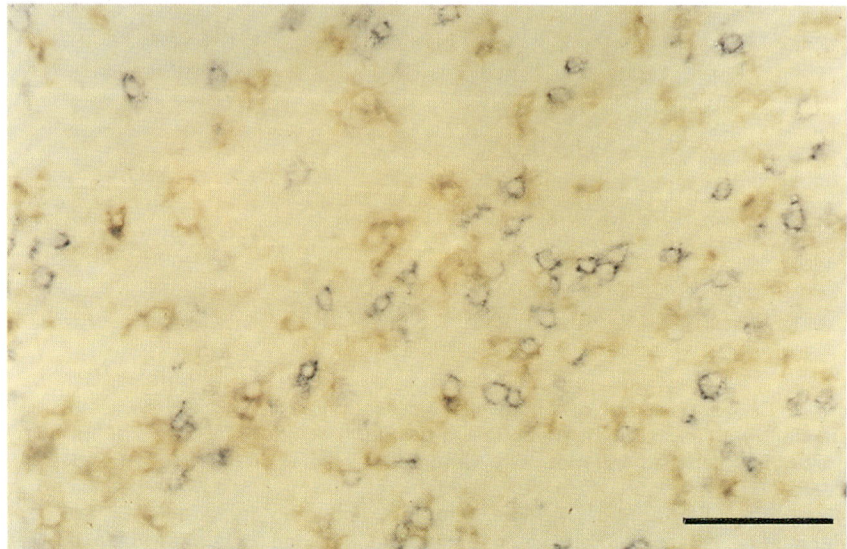

Figure 3. A frozen section of non-pregnant uterine endometrium showing macrophages (brown) localized with Leu M3 antibody in conjunction with the immunoperoxidase technique using diaminobenzidine as chromogen and natural killer cells (blue) localized with NKH1 antibody (CD56) in conjunction with the alkaline phosphatase technique using the chromogen Fast Blue BB as described in *Protocol 4*. (Kindly supplied by Mr P. Jackson, Academic Unit of Pathology, Department of Clinical Medicine, School of Medicine, The University of Leeds, Leeds, UK.) Scale bar 50 μm.

sufficient colour intensity for visualization by normal transmitted light was obtained with these particular applications they may not be universally applicable.

3.2 Immunolabelling techniques

There is a diverse range of microscopically dense probes available for light microscope immunocytochemistry and in consequence there is potentially a wide variety of double immunolabelling techniques that can be performed. A selection is given to show the principles of the major techniques and these can be adapted to suit individual requirements. The schedules are either sequential (*Protocols 1–5*) or simultaneous (*Protocols 6* and *7*). The techniques can be applied to cryostat, wax, and resin sections as well as smears. Pre-treatment depends upon the nature of the specimen. If paraffin sections are used they need to be deparaffinized and hydrated. Resin sections need to be etched and enzyme digestion may be necessary. Frozen sections will need to be fixed (Chapter 3). Phosphate (0.01 M, pH 7.22)-buffered 0.15 M saline (PBS) is the most frequently used buffer for diluting reagents and unless otherwise stated all incubations are at room temperature and 100 μl of each reagent is added to the specimen.

Protocol 1. Indirect two stage method using primary antibodies raised in the same species in combination with peroxidase alone as a probe

The peroxidase in this technique is developed to two contrasting colours (18, 19). Care must be taken to fully develop the first enzyme since unused enzyme may react again in the second reaction yielding spurious signals. The example is given for mouse primary antibodies. For other species alter the reagents accordingly.

1. Pre-treat specimens as necessary (Chapter 3 and Section 3.2 above) and transfer to PBS.

2. Block endogenous peroxidase with 3% H_2O_2 in deionized water for 5 min.

3. Apply normal rabbit serum diluted 1:5 with PBS for 10 min.

4. Incubate in first primary antibody optimally diluted in PBS containing 1% BSA (PBSA) for 1 h.

5. Wash three times, 3 min each wash, in PBSA.

6. Flood with rabbit anti-mouse immunoglobulin conjugated to peroxidase optimally diluted with PBSA.

7. Wash three times, 3 min each wash, in PBSA.

8. Develop the peroxidase enzyme using diaminobenzidene as chromogen (Chapter 3). This may be performed directly on the slide or in a Coplin jar. Allow the reaction to proceed for 5–7 min checking occasionally with a microscope. Terminate prior to the appearance of background staining by washing the slides three times, 5 min each wash, in PBSA.

9. Repeat steps **3–7** above with the second antibody.

10. Develop the peroxidase of the second layer using 4-chloro-1-naphthol (Chapter 3), again with periodic microscope examination, for 5–10 min.

11. Wash three times, 5 min each wash, in PBSA.

12. Mount in an aqueous medium such as Glycergel before viewing.

Protocol 2. Indirect three stage method utilizing avidin–biotin technology

Two contrasting fluorochromes are used in this technique. Any reactivity of the avidin from the first incubation is blocked by an incubation with biotin before application of the second reagents (20).

1. Pre-treat specimens as necessary (Chapter 3 and Section 3.2 above) and transfer to PBS.

Protocol 2. *Continued*

2. Apply first biotin-labelled primary antibody, appropriately diluted in PBS containing 1% BSA (PBSA) for 30 min.

3. Wash three times, 3 min each wash, in PBS.

4. Apply avidin–FITC diluted in PBSA for 30 min.

5. Wash three times, 3 min each wash, in PBS.

6. Apply biotin (100 µg/ml in PBSA) for 15 min.

7. Wash three times, 3 min each wash, in PBS.

8. Repeat steps 2–5 as above, but using a contrasting fluorochrome such as phycoerythrin or rhodamine.

9. Mount in 50% aqueous glycerol or a commercial fade retardant medium and view using a fluorescent microscope equipped with relevant filters.

Protocol 3. Combination technique using immunogold silver staining with immunoenzymatic labelling

This sensitive technique uses a three step (streptavidin–biotin) immunogold silver staining method in conjunction with a three step alkaline phosphatase technique (15). Primary antibodies of the same species are described. For other species alter the reagents accordingly. One antibody is conjugated to FITC to enable spcific double immunolabelling to be performed. Both the silver enhancement and the alkaline phosphatase development should be monitored with a microscope.

1. Pre-treat specimens as necessary (Chapter 3 and Section 3.2 above) and transfer to PBS containing 5% skim milk powder and 0.1% Tween 20 in PBS, pH 7.4 (PST).

2. Incubate with PST containing 20% normal sheep serum for 30 min.

3. Apply first primary monoclonal antibody appropriately diluted in PST overnight at 4 °C.

4. Wash three times, 3 min each wash in PST.

5. Apply anti-mouse Ig conjugated to biotin diluted in PST for 1 h.

6. Wash three times, 3 min each wash, with PST followed by three 3 min washes in PBS containing Tween 20 (0.1%) and BSA (0.2%).

7. Apply streptavidin-conjugated gold (LM grade) diluted in PBS containing 1% BSA for 1 h.

8. Wash three times, 3 min each wash, in PBS.

9. Incubate in PST containing 10% normal mouse serum for 20 min.

10. Apply the second primary antibody (FITC conjugated), appropriately diluted in PST for 2 h.

11. Wash three times, 3 min each wash, in PST.

12. Apply rabbit anti-FITC Ig diluted in PST for 1 h.

13. Wash three times, 3 min each wash, in PST.

14. Apply optimum concentration of sheep anti-rabbit Ig conjugated to alkaline phosphatase diluted with PST for 1 h.

15. Wash three times, 3 min each wash, in PBS containing Tween 20 (0.1%) and BSA (0.2%).

16. Silver enhance (see Chapter 3).

17. Develop the alkaline phosphatase using Fast Blue BB[a] as the chromogen (2–20 min).

18. Mount in an aqueous medium such as Glycergel before viewing.

---

[a] Dissolve 2 mg naphthol AS-MX phosphate in 0.2 ml *N,N*-dimethylformamide in a glass tube add 9.8 ml of 0.1 M tris buffer pH 8.2 and 0.01 ml of 1 M levamisole. This may be stored at 4 °C for several weeks or longer if frozen at −20 °C. Immediately before use dissolve 10 mg Fast Blue BB in the solution and filter. Incubate the sections in the mixture for 10–20 min at room temperature, periodically checking the reaction with a microscope, and stop the reaction by rinsing with distilled water.

Protocol 4. Combination approach using a conventional two step indirect technique in association with an avidin–biotin method

This method (21) is suitable for primary antibodies raised in the same species (in this case mouse) and is both sensitive and reliable.

1. Pre-treat specimens as required (Chapter 3 and Section 3.2 above) and transfer to PBS.

2. Block endogenous biotin if necessary using streptavidin (0.1% in PBS, 30 min) followed by d-biotin (0.01% in PBS, 30 min).

3. Inhibit endogenous peroxidase with 3% H_2O_2 in deionized water for 5 min.

4. Wash twice, 3 min each wash, in PBS.

5. Apply first primary monoclonal antibody appropriately diluted in PBS containing 1% BSA (PBSA) for 1 h.

6. Wash three times, 3 min each wash, in PBSA.

7. Apply goat anti-mouse Ig conjugated to alkaline phosphatase diluted in PBSA for 30 min.

Protocol 4. *Continued*

8. Wash three times, 3 min each wash, in PBSA.

9. Apply undiluted normal mouse serum for 30 min.

10. Incubate in second primary monoclonal antibody conjugated to biotin diluted in PBSA for 1 h.

11. Wash three times, 3 min each wash, in PBSA.

12. Apply peroxidase-conjugated anti-biotin complex diluted with PBSA for 30 min.

13. Wash three times, 5 min each wash, in PBSA.

14. Develop alkaline phosphatase using Fast Blue BB (Protocol 3).

15. Develop peroxidase using diaminobenzidene (Chapter 3).

16. Counterstain and mount in an aqueous medium such as Glycergel before viewing.

Protocol 5. Formaldehyde vapour technique

One antigenic site is immunolabelled and any remaining antigenicity is blocked with hot formaldehyde vapour to reduce unwanted attachment of reagents from the second immunolablling (22). The technique was described for both light and electron microscopy and may be used if at least one of the antigens can withstand the formaldehyde and heat treatment. The technique is very susceptible to unwanted cross reactions and before use the reagents should be carefully titrated to minimize reagent concentration and the specimens should be exposed to minimum formaldehyde vapour. The specimen must be sufficiently exposed to the formaldehyde vapour, however, to block reactive binding sites. The FITC used in the immunolabelling is slightly quenched by the formaldehyde vapour. Therefore if there is a very weak first antibody and the signal is expected to be low, TRITC should be used as the fluorophore for the first primary antibody.

1. Pre-treat specimens as required (Chapter 3 and Section 3.2 above) and transfer to 1% normal serum in Tris (0.02 M)-buffered 0.9% saline pH 7.4 (TBS) for 1 h.

2. Incubate in the first primary antibody diluted in TBS for 20 h at 4 °C and subsequently at room temperature for 1 h.

3. Rinse sections three times, 10 min each rinse, with TBS containing 1% Triton X-100 (TBST).

4. Incubate with appropriate fluorochrome-conjugated antibody diluted with TBS 1:20 for 1 h.

5. Wash sections three times in TBST, 10 min each wash, dehydrate in ascending ethanols, clear in xylene, and air dry.

6. Place the dried sections in a 1 litre jar, containing 3 g paraformaldehyde powder. Close the jar and keep in an oven at 80 °C for 1 h. Take care to avoid escape of the formaldehyde fumes and place the oven in a ventilation hood.

7. Remove the sections from the jar and transfer to TBST and rinse four times, 15 min each rinse.

8. Incubate with TBS for 30 min.

9. Repeat immunolabelling steps **1–4** using a different primary antibody and a different fluorophore.

10. Rinse three times in TBST, 10 min each rinse.

11. Mount in 50% aqueous glycerol or a commercial fade retardant medium and view using a fluorescent microscope equipped with relevant filters.

In *Protocols 1–5* above the risk of unwanted cross reactions is usually high because the reagents used to detect the second antigen may attach to any unused binding sites on the reagents used to detect the first antigen. The blocking techniques given are usually adequate, but should non-specific immunolabelling persist apply one of the following methods after the first immunolabelling schedule to block or saturate any free binding sites (a–e) or to remove or denature the antibody complexes attached to the first antigen (f, g).

(a) Wash sections for 2–14 h in PBS containing 1% BSA and 0.01% Triton X-100 (23).

(b) Wash sections for 20 min in 10% normal serum (same species as primary antibody) in PBS containing 5% skim milk powder and 0.1% Tween 20 (15).

(c) Treat sections for 30 min in undiluted normal serum of the same species as primary antibody (24).

(d) If the immunoperoxidase techniques are being used develop with DAB to detect the first antigen prior to application of the second layer reagents (25).

(e) If immunogold probes are being used, silver enhance the first layer reagents (which effectively inactivate their immunological properties) prior to application of the second layer reagents.

(f) Treat sections with glycine–hydrochloric acid buffer pH 7.2 for 1 h.

(g) Treat sections for 1 h in unbuffered hydrochloric acid pH 2.0 (10).

Protocol 6. Indirect two stage method using primary antibodies raised in different species (mouse and rabbit) in conjunction with fluorescent probes

Since there is very little likelihood of cross reactions between the two sets of reagents, the primary antibodies may be applied simultaneously, followed by a mixture of the secondary reagents, making this a relatively fast technique. It has been used (26) to identify three different antigens using three different antibodies raised in goat, rat, and rabbit, with corresponding anti-species lissamine rhodamine, fluorescein, and DAMC probes.

1. Pre-treat sections as necessary (Chapter 3 and Section 3.2 above) and transfer to PBS.

2. Apply normal sheep serum diluted 1:5 with PBS for 10 min to reduce non-specific background staining. Drain carefully, but do not wash off.

3. Apply a mixture of the primary antibodies diluted with PBS containing 1% BSA (PBSA) for 1 h.

4. Wash three times, 3 min each wash, in PBSA.

5. Apply a mixture of sheep anti-mouse FITC and sheep anti-rabbit TRITC diluted with PBSA for 1 h.

6. Wash three times, 3 min each wash, in PBSA.

7. Mount in 50% aqueous glycerol or a commercial fade retardant medium and view using a fluorescent microscope equipped with relevant filters.

Protocol 7. Labelled antibody technique

Direct labelling with the primary antibody conjugated to the probe is the safest technique available since cross-reactions between reagents are avoided. Primary antibodies, irrespective of their species may be used simultaneously making the procedure very quick, particularly if fluorochromes are employed. However, the technique is relatively inflexible as each individual primary antibody has to be conjugated to an appropriate label and it is inherently less sensitive than direct techniques since covalent binding of the marker to the antibody may result in loss of antibody reactivity. In addition, it is sometimes difficult to obtain sufficient quantities of antibody for conjugation to the dense markers. This protocol describes a useful variant of the technique, utilizing two reagents made by incubating the antibody with the probe before multiple immunolabelling is initiated (27). The technique is described for rat and mouse monoclonal antibodies, although any species of antibody could be used, adjusting the reagents accordingly.

A. *Preparation of the immunological complexes*

1. Dilute the rat monoclonal antibody 10-, 20-, 30-, 40-, and 50-fold in the peroxidase anti-rat Ig to yield to a final dilution of the latter of 1:10, 1:20, 1:30, 1:40, 1:50, and 1:100. Up to 30 dilutions are made with normal serum containing 1% normal sheep serum and 1% bovine serum albumin as the second diluent. Leave for 40 min.

2. Block any remaining secondary antibody binding sites by adding 1% normal rat sera to the complexes. Leave for 20 min.

3. Store the complexes at 4 °C.

4. Prepare mouse antibody–anti-mouse-alkaline phosphatase complexes as in steps **1–3** above, using normal mouse serum at step **2**.

B. *Testing the antibody–enzyme complexes*

1. Pre-treat the specimen (Chapter 3 and Section 3.2 above) and wash the sections three times in PBS, 5 min each wash.

2. Cover specimen with normal serum for 30 min.

3. Drain the normal serum and incubate with dilutions of the pre-formed complexes in PBS for 1 h.

4. Wash three times in PBS, 5 min each wash.

5. *Either* (a) immerse 0.1 M acetate buffer (pH 4.6) for 5 min and develop the peroxidase with 5 ml of 1% 3-amino-9-ethylcarbazole (dissolved in *N,N*-dimethylformamide), 85 ml of 0.1 M acetate buffer (pH 4.6) and 100 µl of hydrogen peroxide for 8 min, checking occasionally with a microscope, *or* (b) immerse in 0.05 M Tris–HCl buffer (pH 8.2) for 5 min and develop the alkaline phosphatase with freshly filtered 20 mg naphthol AS-BI-phosphate dissolved in 200 µl of *N,N*-dimethylformamide, 9.6 mg of levamisole, and 20 mg of Fast Blue BB salt in 40 ml of 0.05 M Tris–HCl buffer (pH 8.2) for 12 min checking occasionally with a light microscope.

6. Check the result for the optimum dilutions of pre-formed complexes.

C. *Double immunolabelling*

1. Prepare the specimen as in B, steps **1** and **2**.

2. Incubate the section with a mixture of the optimally diluted complexes for 1 h.

3. Wash three times in PBS, 5 min each wash.

4. Develop the peroxidase followed by development for alkaline phosphatase as described in B, step **5**.

5. Mount in an aqueous medium such as Glycergel before viewing.

4. Electron microscopy

4.1 Choice of probes

Two microscopically distinguishable probes are needed. Although the enzyme reaction products have been used for electron immunocytochemistry the reaction product is diffuse and may obscure parts of the specimen. Contrast is obtained in the electron microscope by electron density, and therefore the enzyme reaction product cannot be developed to different

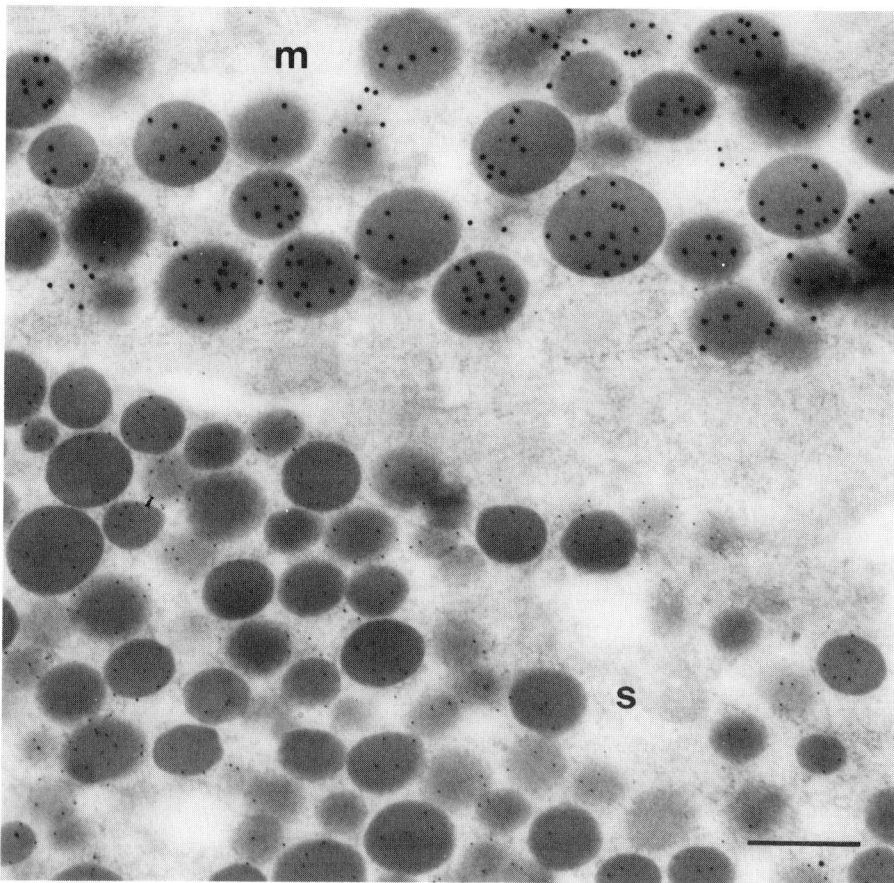

Figure 4. Sections of sheep pituitary tissue, fixed with 4% paraformaldehyde and 0.5% glutaraldehyde and embedded in LR Gold resin, double immunolabelled using the monovalent Protein A–gold technique, *Protocol 9*. The section was immunolablled for prolactin with a 20 nm Protein A–gold probe and for growth hormone with a 10 nm Protein A–gold probe. The micrograph shows part of a mammotroph (m) containing prolactin and a somatotroph (s) containing growth hormone. (Kindly reproduced from Thorpe, J. (1992). *J. Histochem. Cytochem.*, **40**, 435–41.) Scale bar 0.5 μm.

colours as in the light microscope techniques. Although combinations of peroxidase, ferritin, and colloidal gold can be used, the probes currently almost exclusively used for electron microscopy are the colloidal gold probes. Since they can be prepared in several discrete sizes they are ideal for multiple immunolabelling. The sizes commonly used are 5 and 10 nm or 5 and 15 nm probes. For treble immunolabelling all three probes have been used.

There are many single immunolabelling techniques. The pre-embedding, the immunonegative stain, the post-embedding (Chapter 4), and the scanning electron microscope (Chapter 8) techniques are the most widely used. Single immunolabelling with these must be optimized before multiple immuno-labelling is performed. For electron microscopy, combinations of appropriately small colloidal gold particle sizes must be selected to minimize steric hindrance between reagents. As the particle size increases, however, the potential for steric hindrance increases whilst the immunolabelling potential of the probes decreases, making comparison of the relative abundance of each antigen very difficult.

All the protocols described herein employ colloidal gold probes. If the sizes of the gold spheres are not given it is assumed that two contrasting sizes, 5 and 10 or 15 nm, are used. Naturally the type of gold probe used will reflect the nature of the primary antibody. Since small gold probes produce more immunolabelling than larger gold particles immunolabelling obtained with small and large gold probes should not be compared directly. Each antigen

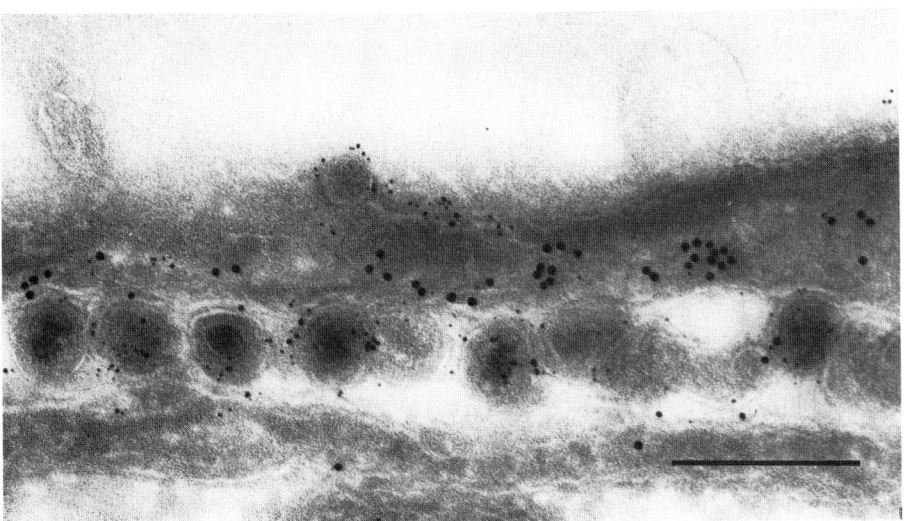

Figure 5. Triple immunolabelling using the denaturation technique *Protocol 10.* Tubulin, glycoprotein, D, and glycoprotein C were localized with 15 nm, 10 nm, and 5 nm colloidal gold probes respectively, in ultrathin frozen sections of human embryonic lung cells infected with Herpes Simples Virus type 1. (Kindly reproduced from Bastholme, L. *et al.,* (1988). *Inst. Phys. Conf. Ser.,* **93**, Vol. 3, Ch. 9, pp. 209–10.) Scale bar 0.3 μm.

should be immunolabelled with each gold probe to confirm immunolabelling efficacy. Examples of multiple immunolabelling for electron microscopy have been presented in Chapter 4, *Figures 9* and *10*, and Chapter 5, *Figures 7* and *8*. *Figures 4–6* herein further demonstrate the versatility of colloidal gold for multiple immunolabelling for electron microscopy.

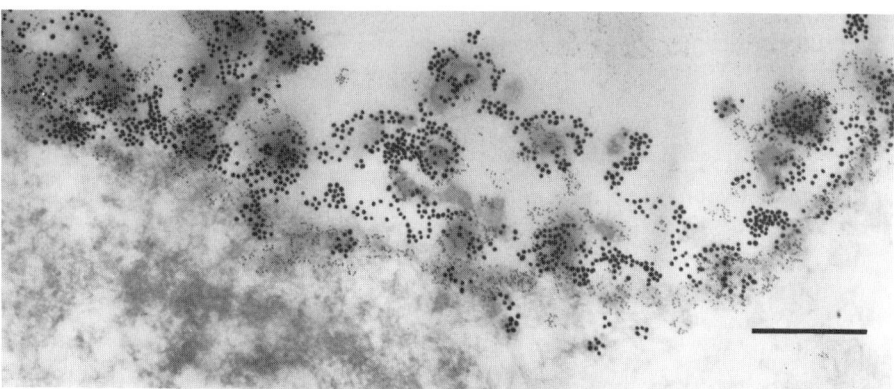

Figure 6. Double immunolabelling using the double-sided technique (*Protocol 12*). Section of infiltrating duct carcinoma of the breast, fixed with 1% paraformaldehyde and embedded in LR White, immunolabelled on one side with antibody against Ca1 and with the 20 nm gold probe and on the other with antibody against HMFG2 and with a 5 nm gold probe. (From a study in collaboration with Drs U. Beckford and S. Chantler, Wellcome Research Laboratories, Beckenham, Kent BR3 3BS, UK.) Scale 0.05 μm.

4.2 Immunolabelling techniques

Electron microscope multiple immunolabelling techniques, like light micro-scope techniques are sequential (*Protocols 8–12*) or simultaneous (*Protocols 13* and *14*). The methods are commonly described for the post-embedding techniques although the antigen–gold (*Protocol 8*), the Protein A–gold (*Protocol 9*), the silver enhancement technique (*Protocol 10*), the denaturing technique (*Protocol 11*), the antibody–gold technique (*Protocol 13*), and the multiple-antibody technique (*Protocol 14*) can theoretically be used on any sample whereas the double-sided technique (*Protocol 12*) can only be used for post-embedding multiple immunolabelling of resin sections on uncoated electron microscope grids since it is the only specimen preparation technique where the specimen can be mounted on an uncoated electron microscope grid and is therefore exposed for immunolabelling on both sides.

All dilutions are made with phosphate (0.01 M, pH 7.2)-buffered 0.15 M saline (PBS). It is most important to perform control incubations as described in Section 2.5.

Protocol 8. The antigen−gold technique

This highly specific technique was described by Larsson (28) based on the divalency of IgG molecules in conjunction with antigen-coated colloidal gold spheres (gold-labelled antigen detection technique). The technique is useful for multiple immunolabelling provided the antigen can be purified. Briefly, if antibody is applied in excess, one antigen binding site of the IgG molecule binds to the antigen leaving the other free. This free site is specific for an identical antigen. In this case the antigen−gold probe is used as a microscopically dense marker. These antigen−gold probes are made in a similar fashion to the gold probes described (Chapter 8) and immunolabelling is as follows:

1. Treat the sections with 1:10 dilution of normal serum in PBS for 10 min.
2. Incubate with excess primary antibody diluted in PBS containing 1% BSA (PBSA) for 24 hr at 4 °C.
3. Rinse three times, 5 min each rinse, in PBSA.
4. Expose to antigen−gold probes diluted in PBSA for 2 h.
5. For multiple immunolabelling repeat steps **1−4** with a second antibody and a second sized antigen−gold complex.
6. After washing three times, 5 min each wash, with Tris (0.05 M)-buffered 0.9 saline pH 7.4, post-fix in 1% glutaraldehyde in buffer for 30 min, rinse further in distilled water, and contrast in preparation for viewing with the electron microscope.

Protocol 9. The Protein A−gold technique

This technique is useful for antibodies which can be detected with Protein A. The first antigen is incubated with primary antibody followed by a Protein A−gold probe. Free IgG binding sites on the Protein A−gold are blocked by a further incubation with normal IgG. An IgG molecule can bind more than one Protein A molecule. Free Protein A binding sites on the primary antibody and the normal IgG are blocked by incubation of the specimen with a monovalent fragment of Protein A before the second antigen is localized with the second primary antibody and the second sized Protein A−gold probe (29).

1. Incubate the sections with 2% gelatin in PBS for 10 min.
2. Incubate with 0.02 M glycine in PBS for 10 min to block free aldehyde groups.

Protocol 9. *Continued*

3. Incubate with first primary antibody diluted with PBS for 30 min.

4. Wash four times with PBS, 1 min each wash.

5. Incubate with Protein A–20 nm gold complex diluted with PBS for 1 h.

6. Rinse three times, 10 min each rinse, with PBS.

7. Incubate the sections with undiluted rabbit serum for 2 h.

8. Rinse three times in PBS, 2 min each rinse.

9. Float the sections on 1 mg/ml Protein A fragment (recombinant, mol. wt. 13 885 from *E.coli*) for 1 h.

10. Rinse three times, 2 min each rinse, by floating the grids on PBS.

11. Float the grids on diluted second antibody for 2 h.

12. Float the grids on Protein A–gold probe (10 nm) for 1 h.

13. Rinse three times in PBS, 10 min each rinse.

14. Rinse four times in double glass-distilled water, 3 min each rinse.

15. Contrast by routine methods and examine with the electron microscope.

Protocol 10. The silver enhancement technique

In this method (30) the silver enhancement technique is used to increase the size of the gold marker and to inactivate the anti-species antibody present on the gold sphere. This allows immunolabelling of a second antigen with second primary antibody raised in the same species as the first. The same gold probe is therefore used throughout. It is readily applicable to sections.

1. Block non-specific binding by floating the sections for 5 min on Tris (0.02 M)-buffered 0.9% saline pH 8.2 (TBS) containing 5% normal goat serum and 0.1% bovine serum albumin.

2. Incubate the antigen for 1 h with a suitable concentration of antibody diluted with TBS.

3. Rinse the grids twice, 5 min each rinse, in TBS.

4. Incubate the sections with an anti-species gold probe diluted with TBS for 1 h.

5. Rinse twice, 5 min each rinse, with TBS and jet wash with water.

6. Silver enhance for 4–6 min (Chapter 3) to increase the size of the gold spheres by a factor of two or three.

7. Repeat steps **1–5** with the second antibody and the same gold probe.

8. Fix specimen for 15 min with 2% glutaraldehyde in phosphate buffered saline and prepare for examination with the electron microscope.

Protocol 11. The denaturation technique

A further technique which permits the use of antibodies raised in the same species is the denaturation technique (22). The first antigenic site is localized with the first antibody and gold probe. Free binding sites on this complex are saturated by formaldehyde vapour treatment before immunolabelling with subsequent antibody and further gold probe. This technique has been extended to a triple immunolabelling electron microscope technique (31). The specimen is dried during this procedure and therefore the technique is applicable to the immunonegative stain and the post-embedding techniques. A full technical discussion of this technique has been given for the histological equivalent of this technique (*Protocol 5*).

1. Rinse the sections for 5 min with saline followed by Tris (0.02 M)-buffered 0.9% saline pH 7.4 containing 1% BSA (TBSA) for 5 min.

2. Incubate with a suitable dilution of specific antiserum made with TBSA for 20 h at 4 °C and subsequently 1 h at room temperature.

3. Rinse the sections three times, 10 min each wash, with TBSA.

4. Incubate sections with 5 nm gold probes, diluted with TBSA, for 8 min and jet wash with TBSA, wash three times, 10 min each wash, on droplets of TBSA followed by two 5 min washes in TBS-0.1% triton-X and finally two 5 min washes in distilled water.

5. Air-dry the grids. Place them specimen-side upwards in a 1 litre jar containing 3 g paraformaldehyde powder. Close the jar and place in an oven maintained at 80 °C in a ventilation hood for 1 h.

6. Rinse the grids twice with distilled water, 5 min each wash, followed by two 5 min rinses in TBS-0.1% triton-X and finally rinse in TBSA for 10 min.

7. Immunolabel as in steps 2–4 using a second antibody and a 10 nm gold probe.

8. If a third immunolabelling is required repeat steps 5, 6, and 7 using a third antibody and a 15 nm gold probe.

9. Finally rinse the grids twice in distilled water, 5 min each wash, and contrast with uranyl acetate and lead citrate before examination with the electron microscope.

Protocol 12. The double-sided technique

Cross reactions between reagents can be minimized when post-embedding immunolabelling resin sections by immunolabelling first one side of a section, then immunolabelling the second side taking care that the grid does not sink (32). The author has employed this technique successfully using sections of

Protocol 12. *Continued*

tissue embedded in LR White. A recent report (29) has suggested that although the technique is suitable for sections of tissues embedded in hydrophobic resins, seepage of reagents through sections of tissue embedded in hydrophilic resins such as LR Gold may be a problem.

1. Float the grid, section side down, on a 20 μl droplet of 1% BSA in PBS (PBSA) for 5 min.

2. Float the grid on a 20 μl droplet of PBS containing 1% gelatin for 10 min to reduce non-specific labelling.

3. Float the grid on a 20 μl droplet of 0.02 M glycine in PBS for 3 min, to block remaining free aldehyde groups of the fixative.

4. Rinse the sections with PBSA by sequentially floating the grid on four droplets (20 μl) for 1 min each.

5. Float the grid on a 20 μl droplet of the first primary antibody diluted with PBSA for 1 h.

6. Rinse the sections by floating the grid sequentially on five droplets (20 μl) of PBSA for 1 min each.

7. Float the grid for 1 h on a 20 μl droplet of suitable gold probe diluted with PBSA.

8. Wash the sections by floating the grid on three droplets (20 μl) of PBSA for 1 min each.

9. Wash the sections five times, 1 min each wash, with distilled water and dry in air.

10. Coat the immunolabelled side of the section with a layer of carbon. The carbon film prevents further immunolabelling of antigens on this side of the section.

11. Coat the carboned side of the section with a layer of celloidin to give the section sufficient strength to withstand further immunolabelling and exposure to the electron beam.

12. Immunolabel the reverse uncoated side of the section with antibody and second sized gold probe as described in steps **1–8**.

13. Rinse the sections on five droplets of distilled water for 1 min each.

14. Stain sections with uranyl acetate and lead citrate by routine methods and examine with the electron microscope.

Protocol 13. The antibody–gold technique

A simple and specific technique for multiple immunolabelling is to coat different sized gold particles each with a different antibody (33) (Chapter 8) and use them for immunolabelling. This technique is not often used as the volume of antibody available is usually not sufficient for preparing a gold probe.

1. Block non-specific sites for 15 min with 1% BSA in PBS (PBSA).
2. Incubate the specimen with a mixture of the different antibody–gold complexes at optimal dilution in PBSA for 1 h.
3. Rinse in PBSA and fix briefly with 1% glutaraldehyde in PBS.
4. Rinse with distilled water and contrast as required before examination in the electron microscope.

Protocol 14. The multiple-antibody technique

Different antigens can be conveniently and quickly detected if the reagents used do not cross-react. Immunolabelling is performed by mixing the required dilutions of each reagent and immunolabelling simultaneously (34).

1. Block non-specific labelling with Tris (0.05 M)-buffered 0.9% saline pH 8.2 (TBS) containing 1:30 dilution of normal goat serum for 10 min.
2. Mix suitable concentrations of the two primary antibodies diluted to optimal titres in TBS containing 1% BSA (TBSA) and incubate with the specimen for 1 h.
3. Jet wash the grids with TBS and with TBS containing 0.2% bovine serum albumin.
4. Place the grids on droplets of TBSA for 5 min.
5. Drain the specimens and incubate with a mixture of the required gold probes of different specifications and size, diluted in TBSA for 1 h at room temperature.
6. After jet washing in TBSA, TBS, and distilled water counterstain the specimen as required for electron microscope examination.

---

5. Conclusions

The most common methods of multiple immunolabelling for light and electron microscopy have been described. Multiple immunolabelling is a logical extension of the single immunolabelling technique. Multiple immuno-labelling is more complex than single immunolabelling and there is a high risk of non-specific cross reactions occurring. It is important therefore that each step in the immunolabelling is very carefully controlled.

References

1. Coons, A. H., Creech, H. J., and Jones, R. N. (1941). *Proc Soc. Exp. Biol. Med.*, **47**, 200.
2. Riggs, J. L., Seiwald, R. J., Burckhalter, J. H., Downs, C. M., and Metcalf, C. M. (1958). *Am. J. Pathol.*, **34**, 1081.
3. Hiramoto, R., Engel, K., and Pressman, D. (1958). *Proc. Soc. Exp. Biol.*, **97**, 611.
4. Chadwick, C. S., McEntegart, M. G., and Nairn, R. C. (1958). *The Lancet*, **1**, 412.
5. Titus, J. A., Haugland, R., Sharrow, S. O., and Segal, D. M. (1982). *J. Immunol. Methods*, **50**, 193.
6. Staines, W. A., Yamamoto, T., Daddona, P. E., and Nagy, J. I. (1986). *Neurosci. Lett.*, **70**, 1.
7. Oi, V. T., Glazer, A. N., and Stryer, L. (1982). *J. Cell Biol.*, **93**, 981.
8. Graham, R. C. and Karnovsky, M. J. (1966). *J. Histochem. Cytochem.*, **14**, 291.
9. Graham, R. C., Ludholm. U., and Karnovsky, M. J. (1965). *J. Histochem. Cytochem.*, **13**, 150.
10. Nakane, P. K. (1986). *J. Histochem. Cytochem.*, **16**, 557.
11. Mason, D. Y. and Sammons, R. (1978). *J. Clin. Pathol.*, **31**, 454.
12. Malik, N. J. and Daymon, M. E. (1982). *J. Clin. Pathol.*, **35**, 1092.
13. Bondi, A., Chieregatti, G., Eusebi, V., Fulcheri, E., and Bussolati, G. (1982). *Histochemistry*, **76**, 153.
14. Campbell, G. T. and Bhatnagar, A. S. (1976). *J. Histochem. Cytochem.*, **24**, 448.
15. Gillitzer, R., Berger, R., and Moll, H. (1990). *J. Histochem. Cytochem.*, **38**, 307.
16. Manigley, C. and Roth, J. (1985). *J. Histochem. Cytochem.*, **33**, 1247.
17. Roth, J. (1982). *J. Histochem. Cytochem.*, **30**, 691.
18. Nakane, P. (1968). *J. Histochem. Cytochem.*, **16**, 557.
19. Sternberger, L. A. and Joseph, S. A. (1979). *J. Histochem. Cytochem.*, **27**, 1424.
20. Wood, G. and Warnke, R. J. (1981). *J. Histochem. Cytochem.*, **29**, 1196.
21. Van der Loos, C. M., Van den Oord, J. J., Das, P. K., and Houthoff, H. J. (1988). *Histochem J.*, **20**, 409.
22. Wang, B.-L. and Larsson, L.-I. (1985). *Histochemistry*, **83**, 47.
23. Gown, A. M., Garcia, R., Ferguson, M., Yamanaka, E., and Tippens, D. (1986). *J. Histochem. Cytochem.*, **34**, 403.
24. Van der Loos, C. M., Das, P. K., and Houthoff, H. J. (1987). *J. Histochem. Cytochem.*, **35**, 1199.
25. Sternberger, L. A. and Joseph, S. A. (1979). *J. Histochem. Cytochem.*, **27**, 1424.
26. Staines, W. A., Meister, B., Melander, T., Nagy, J. I., and Hokfelt, T. (1988). *J. Histochem. Cytochem.*, **36**, 145.
27. Krenacs, T., Uda, H., and Tanaka, S. (1991). *J. Histochem. Cytochem.*, **39**, 1719.
28. Larsson, L.-I, (1979). *Nature*, **282**, 743.
29. Thorpe, J. R. (1992). *J. Histochem. Cytochem.*, **40**, 435.
30. Bienz, K., Egger, D., and Pasamontes, L. (1986). *J. Histochem. Cytochem.*, **34**, 1337.
31. Bastholme, L., Nielssen, M. H., Chatterjee, S., and Norrild, B. (1988). *Inst. Phys. Conf. Ser.*, **93** (3), 209.

32. Beesley, J., Beckford, U., and Chantler, S. M. (1984). *Proc. 8th Eur. Cong. E.M.*, **3**, 1599.
33. Lucocq, J. M. and Baschong, W. (1986). *Eur. J. Cell Biol.*, **42**, 332.
34. Tapia, F. J., Varndell, I. M., Probert, L., De Mey, J., and Polak, J. M. (1983). *J. Histochem. Cytochem.*, **31**, 977.

7

Techniques for image analysis

NICHOLAS G. READ and PAULINE C. RHODES

1. Introduction

The types of image produced by the different immunocytochemical procedures described in this book are varied and there is a wide range of equipment and software available for capturing and analysing these images. Computerized image analysis is a powerful tool for assessing immunocytochemical preparations. An objective numerical description of both the morphology and immunobinding in the same sample can be obtained which is not possible with other techniques. Valuable information can be collected on the number, size, geometry, and location of immunolabelled features. In addition, measurements of the intensity of the immunostaining, whether histochemical end product, fluorescence, or immunogold particles can provide an objective means of assessing the specificity of reactions. Such measurements also enable comparisons of the expression of antigenic determinants to be made between different regions of the same sample or between samples. The potential is enormous.

Since much of the procedure can be automated, most or a lot of the subjectivity and user fatigue associated with the qualitative assessment of immunocytochemical staining is eliminated. It also means that the number of samples which can be accurately analysed on a given day is considerably increased.

If meaningful measurements are to be made from immunocytochemical preparations careful consideration must be given to the image analysis system to be used, the experimental design, and the optimization of the immunocytochemical procedure in order to meet certain requirements for image analysis. The purpose of this chapter is to provide the researcher with a guide to the instrumentation and general principles that need to be applied for image analysis of immunocytochemical preparations and to describe in more detail some of the procedures commonly employed.

2. Instrumentation

2.1 Image analysers

There is a wide range of commercially available image analysis systems (1–3).

Some of the instruments have been designed to perform very simple but useful analysis whilst others are highly sophisticated, a factor which is reflected in their price.

There are two main categories of instrumentation, semi-automatic and automatic systems. The terms 'semi-automatic' and 'automatic' seem to mean different things to different people. We are using the term 'semi-automatic image analysis' to refer to procedures where the identification and contouring of immunolabelled objects is carried out interactively by the user. The term 'automatic image analysis' is being used with reference to procedures where feature discrimination is performed automatically.

The simplest semi-automatic system consists of a digitizing tablet linked to a host computer. The microscope image can be projected on to the tablet by a camera lucida or a photograph laid on the digitizer tablet. Objects of interest are traced using a special pen or cross-hair cursor. The X, Y co-ordinates produced by the interaction of the cursor with the digitizing tablet are collected and analysed by the host computer. A more common approach nowadays with semi-automatic systems is to use a video camera linked to the microscope so that the image is displayed on a monitor screen. Interface cards or overlay boards in the computer allow the mixing of the live video image and digitizer signals so that the tracings of objects can be visualized on the monitor. Not all systems employ digitizing tablets but either use a special light pen which interacts with the monitor screen or a mouse device which is familiar to most PC users. The computer software is used to translate the digital co-ordinates, obtained from the tracing, into the specified parameters such as area, perimeter, and maximum diameter. Thus with semi-automatic systems it is possible to obtain information about the geometry and size of immunolabelled features and estimates of the number of immunogold particles associated with particular areas of the sample. The advantage of this approach to image analysis is that it is relatively easy and inexpensive to perform. It also enables images to be analysed which are too complex or of too poor contrast for automatic analysis. The disadvantage is that it can be slow and does not provide spectral or optical density information which is required if the intensity of the immunostaining is to be measured. The latter information can be obtained with automatic image analysis systems (*Figure 1*).

The automatic monochrome image analysis systems exploit grey level differences between various components in the image and the background. The images acquired by the TV camera are digitized (captured) by an analogue/digital converter into a regular array of picture points known as 'pixels'. An array of 512 × 512 pixels is frequently used. This means that the original video image is converted into 262 144 pixels. Each pixel is assigned a number, for example 0 to 255, which represents the grey level, that is the brightness or intensity of the image at that point in the array, where 0 = black and 255 = white. These values are stored in a specific location in

Figure 1. (a) Image analysis system. (b) Key: 1, image analyser (control processor, image processor, image store); 2, optical disk drive (data and image archive); 3, video printer; 4, printer for data output; 5, keyboard; 6, digitizing tablet and mouse; 7, menu monitor; 8, colour image monitor; 9, TV monitor for live image; 10, camera control unit; 11, video camera; 12, light microscope; 13, operator; 14, power filter unit.

the computer memory. Any or all of these pixel intensity values can be amended by software algorithms. These mathematical transformations, or operators, constitute 'image processing' and there are many algorithms that can be applied to the grey image. In some instances, image processing may be used simply to make it easier for the human eye to interpret the digital image without any quantification or analysis taking place. For instance different pseudocolours can be allocated to particular grey levels. When applied to images of immunostained preparations this mapping technique can amplify subtle differences in the intensity of immunolabelling. Image processing is usually necessary before quantification to make the images easier to analyse. It must be kept to a minimum, however, so that essential detail in the images is not removed. The aim of any image processing should be to produce an image which is unambiguous and of biological relevance.

Automatic colour systems, in addition to that described above for the monochrome systems, exploit either the RGB (red, green, blue) or HSI (hue, saturation, intensity) components in the images for the automatic discrimination of features of interest. It should be noted, however, that colour cameras are not as sensitive as monochrome cameras and may limit the resolution of certain analytical procedures. Also, because the analyser is required to deal with a lot more information the processing speed is reduced. Therefore, before undertaking real colour image analysis it is worth considering whether another approach can be adopted. For instance, different coloured components in an image can be analysed on monochrome analysers by using appropriate narrow band filters.

Where the areas of interest within the sample are readily discriminated by the automatic image analysis system the whole analysis procedure can be automated, thereby further reducing the amount of user interaction required and hence the subjective element present in the semi-automatic analysis. Where the microscope is fitted with a motorized stage and automatic focus facilities the analyser can be programmed to drive the microscope. Such situations, however, are rare since most image analysis procedures will require some degree of user interaction.

The basic hardware and software requirements for quantitative immuno-cytochemistry, enabling densitometric measurements as well as the automatic recognition and extraction of labelled features, have been described elsewhere (3). It is important that the system has adequate spatial and grey scale resolution and appropriate image processing capabilities. For most immuno-cytochemistry applications a spatial resolution of 512 × 512 pixels is usually adequate and an 8-bit (256) grey scale resolution is desirable; however, a 6-bit (64) grey scale resolution is adequate. Of course it should be borne in mind that the camera will have a limited resolving power.

When considering purchasing a computerized image analysis system it is important to think of both present and future requirements and whether the system may need to be expanded at a later date. Another important consideration is that if the capabilities of the larger, more versatile image analysis systems are to be exploited fully a dedicated user may be needed. This person would be responsible for developing the analysis procedures and overseeing and advising other users. Other important considerations when purchasing an image analyser include the image processor speed, image analysis capabilities, and image memory size. For instance a single, monochrome, 512 × 512 pixel image with 8 bits of grey level information takes 256 Kbytes to store it. Whatever you do, make sure the machines do what the manufacturers claim they do. It will nearly always be necessary to make modifications, albeit minor, to either the hardware or software. Therefore, it is important to make sure that the supplier will provide good technical and applications back up. The best people to approach for advice are the current users.

It is possible to build your own system and there are a number of papers which describe the criteria for effective hardware selection and software design (4). Whatever system you purchase or build you may find that the data processing facilities do not meet all your requirements. It is advantageous therefore to adopt industry standard file formats such as 'ASCII' files so that data can be transferred to other systems for storage and possible further statistical analysis.

2.2 Cameras

Careful consideration should be given to the choice of input device since this can have a fundamental influence on the performance of the image analysis system. There are two main types of camera, the thermionic-tube cameras such as Vidicon, Plumbicon, Newvicon, and Chalnicon and the solid-state devices which include the charge-coupled devices (CCD) and charge-injection devices (CID). The choice of the camera will be determined by the application for which it is to be used and its cost. The most demanding applications, with reference to immunocytochemistry, are densitometry and low light imaging such as fluorescence microscopy. Until recently the thermionic-tube cameras have usually been used for these applications. In particular the Plumbicon and Newvicon cameras have been most favoured because of the sensitivity and linearity of response (5). However, the performance of the solid-state cameras has improved markedly in recent years and the cameras now have significant advantages over the tube cameras. For instance the precise positioning of their photosensors means that image distortion is minimized. This is obviously a very important feature in image analysis. They also do not suffer from image lag or burning due to high illumination, they are insensitive to magnetic interference, and possess high stability and high sensitivity to light (6). CCD cameras are therefore now being increasingly used for densitometry. Also the image-intensified, cooled CCD cameras are ideally suited for low light level applications including fluorescence microscopy.

As already mentioned, colour cameras have lower sensitivity than monochrome cameras. If colour image analysis is a major requirement, and it cannot be performed using a monochrome camera together with narrow-band filters, more than one camera may be needed. It is also important to note that many cameras are fitted with automatic gain controls which compensate for changes in incident light by adjusting their output to a pre-set level. If densitometric measurements are to be made this automatic gain circuit must be disabled.

2.3 Microscopes

Careful consideration needs to be given to the microscopes, as these provide the original image.

Light microscopes needs to be fitted with good quality, flat field objective lenses such as planachromats or planapochromats. Dark field, phase, and differential interference contrast images should not be used to make dimensional measurements because the edge produced in these images does not necessarily coincide with the true edge. These images can be useful, however, for counting the number of objects in a field. Fitting a continuous interference filter monochromator will help improve the contrast.

If optical density measurements are to be made it is important that a stabilized light source is used. The light intensity will need to be monitored and adjusted to ensure that it remains constant during the analysis procedure. Small differences in voltage can have profound effects on such measurements. Both the microscope and video camera may need to be protected against voltage fluctuations with constant voltage sinusoidal transformers.

3. Basic steps

The three basic requirements for obtaining meaningful results from the image analysis of immunocytochemical preparations are:

- planning
- getting the best starting image
- getting the best from the image analysis system

3.1 Planning

Whenever image analysis is to be employed its specific requirements need to be taken into account when planning the experiment. The inclusion of image analysis as an afterthought should be avoided. Firstly, a clear understanding of the biological process or feature to be investigated needs to be obtained and a decision made as to what would be meaningful to measure. One needs to be critical about the type of measurements to be made otherwise a lot of time can be wasted measuring things of very little relevance to the investigation. There are two main categories of measurement for immunocyto-chemical investigations. The first category includes the topographical and morphometric type of measurement where the number, size, geometry, and spatial relationship of features of interest are assessed. The second category includes densitometric measurements which can give an indication of the density of immunoreaction product and information about variations in antigen concentration. It is, however, extremely difficult to achieve absolute measurements from immunocytochemical preparations by image analysis. This is due to the large number of variables associated with the preparative procedures, including the immunocytochemistry, as well as the actual image analysis. Therefore, whenever possible, relative measurements should be made such as comparisons between cells or features in the same section, or of test sections with controls which have been processed at the same time. Thus measurements of the relative increase or decrease of the morphometric parameters or antigen levels are obtained.

Having decided what will be measured a decision is needed on the number of samples to be analysed for meaningful results. Two important factors which will influence this decision are the expected size of the effect and the variability between samples. The greater the variability and/or the smaller the effect, the greater the number of samples that will be needed. Pilot studies should be performed to determine an appropriate sample size and to help identify the major sources of error in the whole experimental procedure, including the analysis. Both sampling and specimen preparation can introduce large errors. Sampling procedures should ensure that bias is not introduced and the samples are representative of the bulk of the material. The stereologists have examined this problem extensively (7).

The image analysis procedure needs to be carefully considered and any variation introduced by user subjectivity randomized. Specimens for analysis need to be coded and read in a random fashion without knowledge of their treatment. Care must be taken that bias is not introduced into the selection of areas to be analysed. Therefore, a scheme needs to be devised and adhered to in the selection of the areas and the magnification to be employed. Most analysis procedures, no matter how automated, will still require some degree of user control. Ideally for each set of preparations, all the slides or electron microsope (EM) grids should be read on the same day by the same user. Appropriate methods of calibration for both the geometric and optical density measurements need to be developed. Also controls should be included to assess the accuracy of the measuring program.

At the end of the day you need to ask the question 'am I really measuring what I think I am measuring?'. For instance when interpreting the morphometric measurements, made by the image analyser, it must be remembered that they are only giving two dimensional information. Their translation into three dimensional information is full of pitfalls. Therefore, you may need to look at how well the morphometry matches the classical stereological procedures for measuring the functional compartment you are interested in. Densitometric measurements need to be validated by performing parallel biochemical assays such as radio-immunoassay (RIA) as well as microdensitometry or microfluorimetry. Some caution is required, however, when comparing results from different assay procedures. For instance the use of RIA to validate quantitative immunocytochemistry may be inappropriate in many situations since there are differences between the two procedures in the dynamics of antibody binding and in the way the tissues are prepared for analysis (8).

Finally you need to ask the question 'could the same answer be obtained more easily in another way?'. It is important to remember that although very useful information can be obtained by computerized image analysis its use cannot always be justified. You may find that another procedure such as sterology, microdensitometry, and microfluorimetry or a qualitative/subjective assessment may suffice.

3.2 Optimizing the starting image

The computing maxim 'Garbage in, garbage out' or 'GIGO' applies equally well to image analysis. Image processing should never be regarded simply as a method for enhancing poor images. Therefore every effort should be made to obtain the best possible starting image. The quality of the starting image is dependent upon specimen processing, immunocytochemistry and on the optimal operation of the imaging equipment (5, 6). Both the microscope and camera can introduce errors.

3.2.1 The immunocytochemical procedure

The type of measurement to be undertaken will have considerable bearing on the choice of immunocytochemical procedure. If comparisons are to be made between specimens it is essential that they are handled identically, therefore protocols must be strictly followed. Ideally the specimens should be processed at the same time as a single batch, using the same fixatives and immunoreagents. Section thickness and quality need to be carefully controlled. If comparisons are to be made between batches, internal standards should be included with each batch. Different batches of immunoreagents should be tested against standards using several antibody dilutions to ensure that the reagents are equivalent.

If only morphometric measurements are to be made, a relatively high concentration of antibody should be used to ensure that every feature or area, including those with low antigen content, are discriminated. If the aim, however, is to identify variations in levels of antigen, the primary antibody needs to be precisely diluted to achieve a linear relationship between the intensity of immunostaining and the concentration of immunoassayable antigen (8, 9).

The image analyser can be employed to measure the mean grey level of the immunostained area which enables comparisons to be made concerning the relative staining intensity of different areas in a sample or between samples. If the analyser is calibrated against neutral density filters, measurements can be expressed as optical densities. Finally, attempts at estimating the concentration of the antigen can be made if appropriate standards are devised which allow optical density to be calibrated against known antigen concentrations (8). The use of such standards does have certain pitfalls. For instance, their use assumes that the reactivity of antigen in the standard is the same as the tissue antigen and that the antiserum does not cross-react with other antigens in the tissue. Increasing numbers of investigators are now using densitometry to measure levels of immunoreactive agent in tissue sections (8–11). The peroxidase–anti-peroxidase (PAP) technique shows great potential for densitometry studies and the optimal conditions required to obtain meaningful measurements are well established (12). Pilot studies need to be performed with each new application and the resolution of the procedures assessed. For instance, when employing the PAP technique, three major variables need to be optimized. These include the antibody dilution, the concentration of diaminobenzidine (DAB), and the time of incubation in DAB. These variables need to be adjusted so that within a defined range of antigen concentrations there is a linear relationship between intensity of immunostaining and the concentration of immunoreactive agent. The optimal reaction conditions are established for these variables by repeating the immunocytochemical procedure three times and each time one variable is manipulated and the other two are constant (9).

An alternative method to densitometry for detecting differences in antigen levels is the 'supraoptimal dilution technique' (13, 14). This technique is less time consuming than densitometric techniques, it is relatively simple to perform, and is an indirect method for providing semi-quantitative estimates of the differences in antigen levels. It employs both optimal concentrations of antibody, for maximal staining and demonstration of the maximum number of immunoreactive features, and suboptimal concentrations of antibody for demonstrating differences in antigen levels between features. Where antigen levels are low the staining is reduced or completely abolished when the sub-optimal concentrations of antibody are employed. Thus differences between features with high and low antigen levels are more readily detectable and will be reflected in estimations of the number of immunoreactive objects counted by image analysis. If the dilution of the primary antiserum is increased to a point where all staining is abolished the antiserum concentration at that point can be related to the concentration of antigen. The supraoptimal dilution technique appears to be more sensitive than densitometric techniques in detecting differences in immunoreactivity between experimental groups, but unlike densitometric techniques it cannot estimate the magnitude of such differences (11).

Whatever immunocytochemical procedure is used it is important to remember that the image analyser cannot distinguish between artefact and true staining. Therefore good quality assurance is needed before image analysis is performed. This is particularly important with automatic image analysis procedures where user interaction is to be kept to a minimum.

3.2.2 Microscopy

During the analysis of a set of specimens the microscope setting should be kept constant. It is important that the manufacturer's instructions for complete alignment and optimization are referred to. Transmission electron microscopes (TEM) must be set for optimal performance with the lenses correctly aligned and with an even illumination across the field. Light microscope (LM) optics should be clean and adjusted to Köhler illumination. Incorrectly adjusted diaphragms will affect the contrast and the accuracy of feature measurements. The need for good quality flat field lenses has already been mentioned (Section 2.3). Inferior lenses will produce distortion in the projected image and give rise to errors. This aspect needs to be checked by placing a grid-graticule on the microscope stage and displaying the image on the analyser monitor. The monitor itself might produce distortions. This is checked by viewing a projected grid produced electronically by the image analyser. If bad sectors in either the lens or monitor are identified they must be avoided when performing the analysis. Avoiding the edge of the microscope field during image capture may give a significant improvement with non-flat field lenses.

3.2.3 Camera operation

The influence of electrical noise on the video image can be minimized by careful adjustment of the incident light so that minimum gain control has to be applied to the signal from the camera tube. The greater the gain control the higher the noise level. Therefore it is better to increase the illumination level and keep the gain control low than to try to compensate for lower illumination by increasing the gain. It is important to remember that any automatic gain control should be disabled before performing densitometric measurements. Video cameras are susceptible to induced noise. If this is a problem, it is possible to reduce it by earthing the camera as well as other equipment in the vicinity of the camera (5).

3.2.4 Production of photographs

If the analysis is to be carried out on photographs careful consideration needs to be given to their preparation. Care needs to be taken so that bias is not introduced in the selection of areas being photographed and each set of negatives will need to include an appropriate calibrated reference. When printing the photomicrographs, care must be taken to ensure that the enlarger is critically set and that errors due to paper shrinkage are taken into account to obtain an accurate assessment of the final magnification.

3.3 Optimizing the analysis system

All systems will have certain limitations which will affect the accuracy of any measurements. Those limitations frequently encountered are temporal drift, image noise, errors in geometric measurement, photometric non-uniformity, and non-linear densitometric transfer (2). Therefore it is important to know how to assess the limitations and how to introduce corrective steps into the analysis procedure.

3.3.1 Temporal drift and the introduction of a warm up period

Image analysis systems can show long term variability in the accuracy of measurements. This is thought to be due to a number of possible factors, especially the tungsten light and/or power supply of the microscope or camera. It can be assessed by a series of either geometric or densitometric measurements of the same samples over a period of several hours and comparing the results. If there is a significant difference between measurements, the equipment will need to be switched on for an appropriate length of time to allow the various components of the system to warm up before performing measurements.

3.3.2 Geometric errors and the introduction of calibration and validation procedures

Errors in geometric measurements can be caused by temporal drift, inaccuracies in the measurement algorithms, spatial non-uniformity, noise, and alterations in the microscope settings (2). Therefore appropriate calibration and validation steps must be included in the analysis procedure. Geometric calibration usually involves measuring a known length on a scale so that subsequent analysis results are given in appropriate units. Stage micrometers and carbon grating replicas (Graticules Ltd, Tonbridge TN9 1RN, UK) are used for calibrating systems interfaced to light and electron microscopes respectively. Some systems require that calibration is performed each time the analyser is used, others have the ability to store the values for future use. Validation should be performed routinely at the beginning and end of each piece of analysis to assess the accuracy of the measurement procedure. It is performed either using the same scales used for calibration or with specially designed graticules. The National Physical Laboratory (NPL, Teddington, Middlesex TW11 0LW, UK) have produced a reference stage graticule for the calibration and validation of image analysers used with light microscopes. This graticule can also be used to look for defects in monitors.

3.3.3 Densitometric errors and the introduction of appropriate calibration and validation procedures

When making densitometric measurements, a linear relationship must exist between the intensity of the light input and the output generated by the image analyser. Non-linearities can be introduced by the camera, the analogue/digital converter (ADC) and the algorithm computing the optical density reading. The densitometric transfer function of the system can be tested by using a range of neutral density filters to vary the light intensity. The relationship between the optical density created by each filter and the output reading of the analyser indicates both the linearity of the densitometric transfer and the dynamic range of the system. These factors need to be determined before performing densitometric measurements. At the beginning of each set of measurements the system must be calibrated with one or more neutral density filters and periodic checks made during extended periods of analysis.

3.3.4 Noise averaging

Variations in voltage and other electrical disturbances create noise in the video image which can be reduced by applying a noise averaging algorithm. This function collects a programmed number of consecutive images and determines the mean grey value of each pixel. The resulting image is stored for further processing. The averaging algorithm can be adapted to increase image brightness.

3.3.5 Photometric non-uniformity and the application of shade correction algorithms

Photometric non-uniformity or shading may be produced if the light source and microscope optics are defective or misaligned or the camera target tube shows unevenness in sensitivity. Shade correction algorithms can be implemented to eliminate such problems. One such algorithm consists of two steps. At the first step a blank 'reference' image with no specimen is captured and a matrix of correction coefficients is calculated. At the second step the analyser applies this matrix to the captured specimen image. A shade correction can also be applied to images where there is a variation in sample thickness. Photometric non-uniformity can be identified by using 'look-up tables' and overlaying the blank grey-level images with pseudocolours. In this way each grey value in the blank image is given a particular pseudocolour. The look-up tables can be used on digitized images but are particularly useful when applied to live on-line video images. They can also be used to check the efficiency of a shade-correction routine.

4. Automatic image analysis

There is a range of image processing operators that can be used in automatic image analysis and the reader needs to refer to specialist books for detailed lists and explanations (5, 15, 16). In the following sections (4.1–4.5) brief descriptions of some of the commonly used operators are given.

4.1 Grey level image enhancement

The most frequently used operators for the enhancement of grey level images are 'normalization', 're-scaling', 'allocation' of pseudocolours, and 'edge of enhancement'. For instance a low contrast image contains few grey values. During normalization or 'histogram stretching' the individual pixel values are linearly stretched and expanded over the whole range of available grey level values from 0 (black) to 255 (white) which improves the contrast of the image. Alternatively, the grey levels present in the image can be selectively re-scaled into defined ranges rather than spread evenly over the whole range. The allocation of pseudocolours to pixels with particular grey values can help distinguish very subtle tonal differences.

Finally edge enhancement operators, which compare each pixel with its neighbours' values, help sharpen edges around objects of interest whilst smoothing the background. Care needs to be taken when applying any additional image processing operators as they may affect the accuracy of subsequent geometric measurements.

4.2 Segmentation and production of the binary image

Segmentation separates the features of interest from the background. The procedure results in the formation of a binary image in which the pixels are usually displayed as either black or white depending on whether they are the pixels associated with the features of interest (white) or background pixels (black). There are several ways that this can be achieved. The simplest way is to draw round the objects or areas of interest using the cursor device as in semi-automatic analysis. This, however, can be time consuming and defeats the object of trying to automate the procedure. The alternative is to segment the image automatically by getting the analyser to distinguish features of interest from the background by either using grey level or colour discrimination (*Figure 2*). The way these are performed will vary depending on available software. In grey level discrimination, or thresholding, features of interest are detected by virtue of their grey values. At the beginning of the analysis the user defines a range of selected grey values which best correspond to those found in the stained features of interest and distinguishes them from the background. The analyser automatically identifies the pixels with grey values in this range and displays them as white in the binary image. Any pixels with grey values outside this range are displayed as black.

Real colour discrimination can be carried out in two ways either by exploiting the RGB or HSI components in the images. In RGB discrimination, the analyser is first used to assess the amount of red, green, and blue information present in several areas of immunostaining and the overall minimum and maximum levels of each colour are registered. These values are used for the automatic discrimination of immunostained features in subsequent images. Thresholding may be based purely on the amount of each colour or

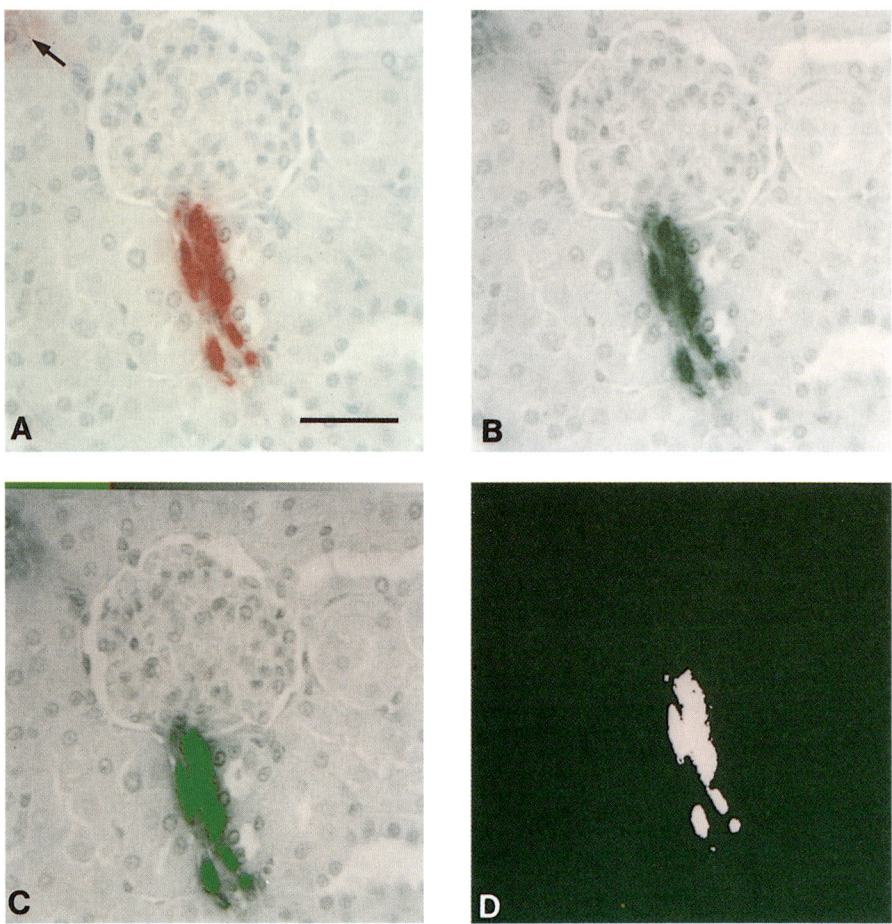

Figure 2. Illustrates some of the key steps performed in an analysis run to discriminate and measure the area of an immunostained juxtaglomerular apparatus (JGA). A: Digitized RGB image of a section of kidney cortex stained for immunoreactive renin using an ABC–AP technique. Strong staining for the JGA and weaker staining of a tubule (arrow) can be seen. Scale bar 50 μm. B: The grey level image of the green component showed greater contrast than either the red or blue components and was therefore used for JGA discrimination. C: Using grey level thresholding the JGA has been discriminated and the weaker staining tubule has been left undiscriminated. The area to be measured has been highlighted with a green overlay. At the top of the picture is a scale of grey values from black to white and the position of the green bar indicates which part of the grey scale is discriminated. D: The final binary image illustrating the area to be measured.

on the combined ratios of colour. In HSI discrimination the analyser recognizes aspects of a colour image in a similar way to the human eye. Hue refers to the colour, saturation to the amount of white light which is mixed with the colour (for example the difference between red and pink), and intensity is the brightness of the colour (similar to grey values). There are various ways these three components can be used to discriminate stained features. The advantage of HSI over RGB is that it can be used to give accurate discrimination even when there are changes in the level of illumination. When using the RGB components in objects to produce binary images for the estimation of their areas, variations in light intensity can have a profound effect on the accuracy of discrimination. The resulting discriminated areas are either too large or too small. When employing HSI, changes in the light level cause little alteration in area detection when hue and saturation (H and S) are treated separately to intensity. The saturation values relate to the intensity of staining, thus weak and strong reactions can be readily distinguished. The actual mechanics of the segmentation with HSI are similar to those of RGB discrimination. Firstly, the analyser is used to assess the levels of H and S in several areas of immunostaining and the overall assessment is registered. Areas of the specimen image are compared with these 'reference' values and a binary image created.

4.3 Object and field selection

The way this is done depends on which facilities are available on the analyser. There are three main options:

(a) The whole field can be measured.

(b) Framing or windowing. This involves the generation of a measuring frame which overlays a portion of the binary image. The use of measuring frames ensures that only whole objects are included in the analysis as objects touching or outside the frame can be identified and eliminated automatically. The type of software available in most analysers enables the operator to determine the size, shape, and position of the measuring frame. Careful thought needs to be given to the type of measuring frame appropriate for the particular application.

(c) Boundary setting. Specific areas for analysis can be determined by outlining them using the cursor device.

4.4 Binary image editing/processing

No matter how successful the thresholding step, most biological samples require some editing to erase unwanted objects or artefacts, to separate overlapping objects, to join fragments of the same object, or to fill in unwanted holes. With some images this will need to be done entirely interactively whilst with others, automatic operators can be used. Interactive image editing allows the operator to use the cursor device like an eraser or paint brush to either add or subtract pixels. There is a range of automatic binary operators available and the reader should consult specialist image analysis books for descriptions of their potential (5, 15, 16). Four binary operators commonly used are scrap, erosion, dilation, and skeletonization. The scrap function can be used to automatically remove objects outside a defined size range. Erosion and dilation operators either subtract or add pixels from the edge of a segmented feature and can be used to separate two objects which are touching or join parts of a single object which have become separated. Skeletonization is a thinning binary operator which reduces objects to a single pixel width and is particularly useful for measuring the length of fibres. The use of binary operators, particularly the latter three, needs to be carefully controlled and the final binary image checked to ensure that important information has not been removed. One final consideration is that if extensive image processing is required it may be more appropriate to perform segmentation interactively rather than automatically.

4.5 Measurement specification

Having produced a binary image of the features of interest, the types of measurements to be made need to be specified.

4.5.1 Counting and geometric measurements

There are two categories of measurement which can be made: field-specific and feature-specific. The most common field-specific measurement is counting, others include total area of the frame, boundary, or field or area fraction and intercept. The most useful feature-specific parameters are area, perimeter, length, breadth, maximum and mimimum diameters, and feret diameters. Additional measurement parameters can be derived from the basic measurements by introducing appropriate arithmetic operators.

4.5.2 Densitometric measurements

Densitometric measurements can be made of an area within an entire frame or boundary or of segmented objects contained within the frame, boundary, or field

The intensity of immunostaining, measured by computerized image analysis, can be expressed in at least three ways:

(a) Grey values. Either as the 'total grey value', which is the sum of the individual pixel values in the area being measured, or the 'mean grey level' within that area.

(b) Optical densities. The sum of the individual pixel values in the area being measured is known as the 'integrated optical density' (IOD). Division of the IOD by the total number of pixels contributing to this yields the 'average optical density' (AOD) per unit area. The AOD gives a measure of the relative concentration of reaction product in the segmented area.

(c) Antigen levels. As already mentioned, if optical density measurements are to be made the system needs to be calibrated against a series of neutral density filters to convert grey values into optical density values (Section 3.3.3). If antigen levels are to be measured, a series of immunostained standards need to be used to produce a calibration curve so that optical density measurements in the specimen images can be related to antigen levels.

5. Image analysis techniques

Immunocytochemical procedures present a whole variety of starting images. Therefore, the way the various measurements are taken will vary with the tissue and immunocytochemical procedure as well as the type of instruments used.

In the following section, outlines of procedures are presented for:

- semi-automatic analysis for counting and geometric measurements (*Protocol 1*)
- automatic analysis for counting and geometric measurements (*Protocol 2*)
- automatic analysis for densitometric measurements (*Protocol 3*)

These outlines can be adapted to satisfy the specific requirements of each investigation. Finally, the analysis of fluorescent (Section 6) and immuno-gold (Section 7) preparations are given further consideration.

Protocol 1. Semi-automatic analysis procedure for counting and geometric measurements

1. Leave the microscope, camera and image analyser switched on for a period of time before making any measurements to allow the system to stabilize (see Section 3.3.1).

2. Select the appropriate objective (LM) or magnification setting (TEM) for the size of objects being measured and ensure that the microscopes are optimally aligned (see Section 3.2.2). Set the brightness of the light source (or electron gun) to match the sensitivity of the camera with minimum gain control (see Section 3.2.3). Where photographs are used great care should be taken in their production (see Section 3.2.4.). Photographs of the appropriate calibration scale (stage micrometer for light micrographs and carbon grating for transmission electron micrographs) must be produced.

3. Calibrate the image analyser using video images or photographs of the stage micrometer (LM) or carbon grating (TEM, see Section 3.3.2).

Protocol 1. *Continued*

4. Enter the parameters to be measured and, where appropriate, arithmetic operators to be computed.

5. Check the accuracy of the procedure by measuring objects of known size. This can be done using either a stage micrometer or NPL graticule for light microscopy, or carbon grating for transmission electron microscopy (see Section 3.3.2).

6. Select areas for analysis taking care not to introduce bias (see Section 3.1).

7. Using the cursor device trace round or count the areas/objects of interest and activate the analyser to compute the measurements.

8. Repeat the procedure outlined in step **5**.

9. Perform appropriate statistical analysis, tabulation, and graphical presentation of data.

Protocol 2. Automatic image analysis procedure for counting and geometric measurements

This procedure can be carried out using video images taken directly from the microscope. The use of photographs should be avoided as it introduces additional sources of error and increases the time and effort required to produce the results (see Section 3.2.4)

1. Leave the microscope, camera, and image analyser switched on for a period of time before performing any measurements to allow the systems to stabilize (see Section 3.3.1).

2. Select the appropriate objective (LM) or magnification (TEM) setting for the size of the objects being measured and ensure that the microscopes are optimally adjusted (see Section 3.2.2). Set the brightness of the light source (or electron gun) to match the sensitivity of the camera (see Section 3.2.3). The intensity of the light source needs to be noted and checked at regular intervals. If a change in the light intensity is detected it must be returned to the original level.

3. Calibrate the image analyser using images of the stage micrometer (LM) or carbon grating (TEM) (see Section 3.3.2).

4. Specify the measurements to be carried out (see Section 4.5).

5. Capture the image of a slightly defocused blank field. This image is used for computing the shade correction matrix which will be applied to each subsequent specimen image (see Section 3.3.5). If the video image is noisy apply a 'noise averaging algorithm' (see Section 3.3.4).

144

6. Check the accuracy of the procedure by measuring objects of known size. This can be done using either the stage micrometers or NPL graticule for light microscopy, or the carbon grating for transmission electron microscopy (see Section 3.3.2).

7. Select areas for analysis taking care not to introduce bias into the selection (see Section 3.1).

8. Check that the image on the monitor screen is focused and capture the image (digitization). If the image is noisy or has a low light intensity apply a 'noise averaging algorithm' (see Section 3.3.4).

9. Apply shade correction to the specimen image (see Section 3.3.5).

10. If necessary perform further image enhancement (see Section 4.1).

11. Segment the image to produce a binary image (see Section 4.2).

12. Select the objects and/or fields to be measured (see Section 4.3).

13. If necessary, perform binary image editing/processing (see Section 4.4).

14. Activate the measurement program (see Section 4.5).

15. Repeat procedure outlined in step **6**.

16. Compute appropriate statistical analysis, tabulation, and graphical presentation of data.

Protocol 3. Automatic analysis procedures for densitometric measurements with light microscope images

The following procedure can be used to obtain both geometric and densitometric data from live light microscope images of specimens stained using techniques, such as PAP and ABC, which produce coloured end products.

1. Leave the microscope, camera, and image analyser switched on for a period of time before performing any measurements to allow the systems to stabilize (see Section 3.3.1).

2. Choose an appropriate objective for the size of objects being measured and ensure that the microscope is optimally aligned (see Section 3.2.2). Select a suitable colour filter or monochromator setting for the analysis procedure. It is important that the same wavelength is used throughout the analysis procedure as absorption is only proportional to mass of adsorbing material when the calibration and measurements are performed under well defined conditions (5).

3. If the camera is fitted with an automatic gain control ensure that this is disabled before carrying out densitometric measurements (see Section 3.2.3.).

Protocol 3. *Continued*

4. Set the brightness of the light source to match the sensitivity of the camera with minimum gain control (see Section 3.2.3). This is done by aligning a blank area of slide in the field of view and adjusting the illumination control and gain control until the peak grey level value is obtained. The intensity of the light source needs to be noted and checked at regular intervals. If a change in the light intensity is detected the above procedure must be repeated.

5. Calibration.

    (a) Geometric: calibrate the image analyser using images of the stage micrometer (see Section 3.3.2).

    (b) Optical density: first measure a totally black field obtained by blocking the light source, thus preventing any light from reaching the camera. **Important**: do not switch off the lamp otherwise this will affect previous settings. Then, using a blank field, insert and measure a series of neutral density filters. Construct a density curve. Most systems have software which will calculate the necessary constants or transformation table for converting grey level measurements into absolute optical density values.

    (c) Antigen versus optical density: using a series of immunostained antigen standards, measure their optical density and produce the necessary constants or transformation table for converting optical density to antigen levels.

6. Capture the image of a slightly defocused blank field. This image is used for computing the shade correction matrix which will be applied to each subsequent specimen image (see Section 3.3.5). If the video image is noisy apply a 'noise averaging algorithm' (see Section 3.3.4).

7. The accuracy of the procedure for performing geometric measurements needs to be checked by measuring objects of known size. This can be done using either the stage micrometers or NPL graticule for light microscopy, or the carbon grating for transmission electron microscopy (see Section 3.3.2). Densitometric measurements need to be validated using neutral-density filters (see Section 3.3.3) and, where appropriate, stained using antigen standards.

8. Select areas for analysis taking care not to introduce bias into the selection (see Section 3.1).

9. Check that the image on the monitor screen is focused and capture the image. If the image is noisy or has a low light intensity apply a 'noise averaging algorithm' (see Section 3.3.4).

10. Apply shade correction to the specimen image (see Section 3.3.5).

11. If required, perform further image enhancement (see Section 4.1). Care

needs to be taken when applying any additional image processing operators as they may affect the accuracy of subsequent densitometric measurements.

12. Segment the image to produce a binary image (see Section 4.2). This step will not be necessary if you choose to perform only densitometric measurements of the entire field, areas included in a measuring frame or features/areas identified using the cursor device. Alternatively a binary image can be produced (see Section 4.2) and used as a template to identify the pixels in the digitized image to be used in the densitometric measurements.

13. Select the objects and fields to be measured (see Section 4.3).

14. If required perform binary image editing/processing (see Section 4.4).

15. Activate the measurement program.

16. When performing densitometric measurements it will be necessary to acquire an image of an unstained area in the tissue section to correct for the contribution of background. The procedure for the image acquisition and densitometric measurements of the selected areas will be the same as described in steps **8** to **14** for the immunostained features/areas. The average grey value or optical density of unstained tissue can then be subtracted from that of the immunostained features/areas of interest and the value multiplied by their area to calculate the total grey value or integrated optical density of immunostaining.

17. Repeat procedure outlined in step **7**.

18. Compute appropriate statistical analysis, tabulation, and graphical presentation of data.

---

6. Measurement of fluorescence intensity or brightness

The inherent sensitivity associated with emission as opposed to absorption processes means that much lower concentrations of immunoreactive agent can be quantified by fluorescence procedures (17, 18).

The procedure employed is similar to that detailed in *Protocol 3*. The value of each pixel is proportional to the fluorescence intensity or brightness at that location. If appropriate standards are employed and calibration curves obtained, intensity measurements can be related to the relative or absolute levels of antigen. Permanent reference standards, such as uranyl glass, are available and which consist of rare earth ions in glass. Their fluorescence is reproducible under a given set of physical conditions and does not fade with exposure. Therefore, they provide a useful standard for calibrating and checking the performance of an image analysis system when measuring fluorescence. Two major complications can arise in quantitative fluorescence microscopy. One is photobleaching of the fluorophores and the other is out-of-plane contributions to the fluorescence intensities. Photobleaching is further complicated since the rate of bleaching can vary in different regions of cells. It is important when such complications do arise that appropriate modifications are made to the procedures otherwise meaningful results will not be obtained (18). Confocal microscopes can be used to reject the light that does not emanate from the plane of focus and when used with an image analyser presents the researcher with a powerful tool for performing micro-fluorimetry.

7. Image analysis of colloidal gold for electron microscopy

The introduction of particulate immunomarkers such as ferritin, iron dextran, and colloidal gold has presented the investigator with the possibility of quantitation of antigens at the electron microscope level (19). Particle counts in combination with area measurements will give a measure of the intensity of the immunoreaction. Counting and area measurement can be performed either using semi-automatic or automatic image analysis procedures similar to those outlined in *Protocols 1* and *2*. The well defined size of the immunogold labels makes them particularly useful for automatic analysis. The correct image magnification must be selected to ensure that the gold particles are clearly discriminated. Comparison of the staining intensities associated with various features in the same section or between sections will give an indication of relative differences in antigen levels. To estimate the actual levels of antigen a knowledge of the relationship between the intensity of immunoreaction and the concentration of the antigen is required. This relationship is frequently referred to as the labelling efficiency, which can be determined using references containing known concentrations of antigen (20). The concentration of antigen in the specimen is determined using the equation:

$$[Ag]s = [Ag]r \times (LDs/LDr)$$

where Ag = antigen concentration, s = specimen, r = reference, and LD = labelling density (counts/unit area).

8. Concluding remarks

Image analysis of immunocytochemical reactions for either light or electron microscopy represents a relatively untapped resource to the researcher. The development of meaningful standards introduces the possibility of obtaining estimates of antigen concentration and spatial resolution unobtainable with biochemical analysis. The hardware and software are being improved continually and are becoming available at much more reasonable prices. The combination of image analysis and confocal microscopy in particular provides us with a powerful procedure for use with both living and processed cells and tissues. Therefore we can expect to see these techniques making a major contribution to our understanding of cell biology.

References

1. Bradbury, S. (1988). *Microscopy*, **36**, 23.
2. Mize, R. R. (1984). *Int. Rev. Cytol.*, **90**, 83.
3. Mize, R. R., Holdefer, R. N., and Nabors, L. B. (1988). *J. Neurosci. Methods*, **26**, 1.
4. Jarvis, L. R. (1986). *Anal. Quant. Cytol. Histol.*, **8**, 201.
5. Joyce-Loebl Ltd (1989). *Image analysis: Principles and Practice*. Short Run Press, Exeter.
6. Jarvis, L. R. (1988). *J. Microsc.*, **150**, 83.
7. Cruz-Orive, L. M. and Weibel, E. R. (1981). *J. Neurosci. Methods*, **122**, 235.
8. Nabors, L. B., Songu-Mize, E., and Mize, R. R. (1988). *J. Microsc.*, **26**, 25.
9. Benno, R. H., Tucker, L. W., Joh, T. H., and Reis, D. J. (1982). *Brain Res.*, **246**, 225.
10. Rahier, J., Stevens, M., De Menten, Y., and Henquin, J-C. (1989). *Lab. Invest.*, **61**, 357.
11. McBride, J. T., Springall, D. R., Winter, R. J. D., and Polak, J. M. (1990). *Am. J. Respir. Cell Mol. Biol.*, **3**, 587.
12. Benno, R. H., Tucker, L. W., Joh, T. H., and Reis, D. J. (1982). *Brain Res.*, **246**, 237.
13. Vacca-Gallaway, L. L. (1985). *Histochemistry*, **83**, 561.
14. Springall, D. R., Collina, G., Barer, G., Suggett, A. J., Bee, D., and Polak, J. M. (1988). *J. Pathol.*, **155**, 259.
15. Gonzalez, R. C. and Wintz, P. (1987). *Digital Image Processing*, Addison-Wesley, Reading, Massachusetts.
16. Russ, J. C. (1990). *Computer-Assisted Microscopy: The Measurement and Analysis of Images*. Plenum, New York.
17. Arndt-Jovin, D. J., Robert-Nicoud, M., Kaufman, S. J., and Jovin, T. M. (1985). *Science*, **230**, 247.
18. Smith, L. C., Benson, D. M., Gotto, A. M., and Bryan, J. (1986). In *Methods in Enzymology*, Vol. 129, (ed. J. J. Albers and J. P. Segrest), pp. 857–73. Academic, London.
19. Griffiths, G. and Hoppeler, H. (1986). *J. Histochem. Cytochem.*, **34**, 1389.
20. Posthuma, G., Slot, J. W., Veenendaal, T., and Geuze, H. J. (1988). *Eur. J. Cell Biol.*, **46**, 327.

8

Correlative video-enhanced light microscopy, high voltage transmission electron microscopy, and field emission scanning electron microscopy for the localization of colloidal gold labels

RALPH M. ALBRECHT, SCOTT R. SIMMONS,
and JAMES B. PAWLEY

1. Introduction

Several chapters in this book have discussed the rationale and methods of using colloidal gold particles conjugated to a variety of biological macro-molecules as labels for observation by either light or electron microscopy. This chapter will discuss the use of colloidal gold labels in correlative studies employing video-enhanced differential interference contrast light microscopy (VDIC), low voltage high resolution scanning electron microscopy (LVSEM), and high voltage transmission electron microscopy (HVEM). This correlative microscopic approach permits investigation of processes in living cells by light microscopy followed by preparation of the specimens for electron microscopy and subsequent examination by scanning and transmission electron microscopy of the exact same cells and cell structures observed in the light microscope (1–3). Features such as the overall distribution of different receptor types on the cell surface, as well as their direction and rate of movement, can be followed in the living cell. Changes in cell structure and receptor distribution can also be observed (3–7). Use of double or triple labelling procedures with several sizes of gold particles conjugated to different ligands or antibodies directed against different surface receptor antigens allows visualization of the positions of surface glycoproteins and other antigenic species relative to one another and permits monitoring changes in these relationships over time (6, 8, 9). While individual gold particles 10 nm to 30 nm in size are well below the resolution limit of the light microscope, they can be detected by their inflated diffraction image in the VDIC (7, 10–13). For confirmation of the identity of gold labels which have been followed on live cells by light microscopy, as well as for determining the relation of the positions of the labels to ultrastructural features of the cell, the specimen can be fixed by chemical or physical procedures while still being observed in the light microscope and prepared for electron microscopy. Examination of whole mount preparations of cells by HVEM can be used to examine the three dimensional architecture of the interior structure of the cell, showing the positions of internalized gold labels and allowing correlation of the positions of gold labels either on the cell surface or inside the cell with internal ultrastructural features (1, 14, 15). Subsequent examination of the specimen by LVSEM provides high resolution images of the cell surface. The final positions of the gold labels on the outer membrane of the cell can be determined in relation to cell surface features at molecular levels of resolution and correlated with information previously obtained by VDIC and HVEM (2, 9, 13). It is important to note that because cell morphology and overall distribution of receptors or surface antigens can be examined on individual cells, differences within a population can be determined. Non-morphological studies often rely on averages of many cells that may not all be at the same stage of development or that may consist of several sub-populations. Comparison of the results of these studies with the unique information arising from correlative microscopic studies often leads to new insights regarding cellular processes (16). The correlative aspect of these techniques is extremely important to the justification of the approach to microscopy described in this chapter. It is therefore useful to present examples to demonstrate the potential of the methodology.

2. Preparation of colloidal gold probes

2.1 Preparation of colloidal gold spheres

Colloidal gold sols may be prepared in particle sizes ranging from 1 to 500 nm. A dilute solution of gold chloride is reduced by one of a variety of reducing agents, the size of the resulting particles depending on the amount and type of agent used. Because all of the gold is reduced in all of the preparations, the resulting particle size is determined by the number of nuclei formed in the initial stages of the reduction procedure. At higher concentrations of reducing agent the number of nuclei formed is high and the gold will crystallize on many centres. Because the total amount of Au is limited the nuclei cannot grow very large. As the reducing agent concentration decreases, fewer nuclei form and the resulting particles are larger (17, 18). Gold sols from 3 nm to 16 nm average diameter can be prepared by the reduction of a dilute solution of $HAuCl_4$ by a mixture of sodium citrate and tannic acid (19). The amount of sodium citrate is held constant and the amount of tannic acid and K_2CO_3 is progressively reduced to produce particles of increasing size. Gold sols produced by this method have a very narrow range of particle sizes; however, a residue of polymerized tannic acid remains which may interfere with the conjugation to some proteins. This residue can be removed by the inclusion of 0.1 to 0.2% hydrogen peroxide (19). Colloidal gold of 5 nm diameter can be prepared by the citrate/tannic acid method, or can be prepared by the reduction of $HAuCl_4$ by phosphorus (20). While this second method produces particles which are slightly more variable in size, it avoids the problem of the tannic acid residue. Gold particles 12–150 nm in diameter are prepared by the reduction of $HAuCl_4$ by trisodium citrate (21). Again, increasing the quantity of citrate added results in a decrease in the size of the particles produced.

Gold chloride is very hygroscopic. It can be bought in small ampules and dissolved all at once or stored over desiccant and weighed into the container in which it will be dissolved. The water used to dissolve it should be distilled and filtered before adding it to the gold. All other solutions introduced to the flask, with the exception of the phosphorus/ether used in the preparation of 5 nm gold, should be filtered through a microporous filter (0.2–0.4 µm pore size) immediately before use. All glassware should be cleaned with a non-ionic detergent and rinsed well with distilled water before use because the presence of small contaminants will affect the formation of nuclei and hence adversely affect the desired size range of the sol. The boiling flask in which the reduction takes place will acquire a coating of gold which will help prevent nucleation of the gold on the glass surface. This flask should be reserved for this purpose alone, and care should be taken not to scratch the inner surface in cleaning. The gold colloids have an indefinite shelf life, and are stored at 4 °C until used.

Protocol 1. The tannic acid/citrate method for production of gold sols ranging from 3 to 16 nm

1. Prepare solution A (80 ml) which consists of 1 ml 1% $HAuCl_4$ in 79 ml distilled water.

2. Prepare solution B (20 ml) which consists of 4 ml 1% trisodium citrate, x ml 1% tannic acid (low molecular weight galloyl glucose, #8835 Mallin-krodt, St Louis, Missouri), and x ml of 25 mM K_2CO_3. When less than 0.5 ml of tannic acid is used, it is not necessary to add the corresponding amount of K_2CO_3 to neutralize the pH. The particle size varies with the amount of tannic acid used.

Protocol 1. *Continued*

Tannic acid (ml)	Particle size (nm)
5	3
2	4
0.5	6
0.12	8
0.07	10
0.01	16

   The solution is brought to a total of 20 ml with distilled water.

3. Warm solutions A and B to 60 °C and mix rapidly. Heat the mixture to boiling to complete the reaction.

Protocol 2. The phosphorus method for producing 5 nm gold sols

1. Mix 240 ml of distilled water with 3 ml of 1% $HAuCl_4$ and neutralize the solution with 0.2 M K_2CO_3.

2. Mix 1 ml of white phosphorus-saturated diethyl ether with an additional 4 ml of ether, and add 2 ml of this to the gold chloride solution prepared in step **1**. Phosphorus is both toxic and flammable and care should be exercised in its handling. Spills and excess solution can be neutralized with a solution of $CuSO_4$.

3. Gently stir mixture at room temperature for 15 min, then heat it to boiling and reflux it for 5 min.

Protocol 3. The citrate method for producing 12 to 30 nm gold sols

1. Add 2 ml of 1% $HAuCl_4$ to 200 ml of distilled water and heat to boiling.

2. Varying amounts (see below) of 1% trisodium citrate should be added with mixing, the solution brought to a boil and refluxed for 30 min.

1% sodium citrate (ml)	Particle size (nm)
5	12
4	16
3	24
2.8	30

3. As the gold chloride is reduced, the solution turns first to black and then to red in colour. The smaller sizes of particle produce an orange-red colour in the sol. Larger particles lend a purple cast to the colour.

2.2 Preparation of protein–gold conjugates

Colloidal gold particles are maintained in solution by electrostatic repulsion. Individual particles carry a net negative charge resulting from the [AuCl$_2$] complex, which is believed to be present on the surface of the particle. Addition of electrolytes to the solution effectively compresses the ionic layer surrounding each particle, allowing particles to approach each other until they are close enough for the short range London–van der Waals forces to be greater than the charge-dependent electrostatic repulsive forces and the particles flocculate and fall out of suspension.

Gold sols can be protected from the flocculating effects of electrolytes by coating them with a stabilizing layer of macromolecules such as proteins, which sterically prevent the particles from approaching each other closely enough to flocculate due to London–van der Waals forces. The adsorption of proteins to gold particles is pH dependent. The net charge of proteins in solution is a function of pH and is neutral at the isoelectric point, where the protein behaves as a zwitterion. Because the protein is least soluble at this point it is only weakly hydrated and conditions are most favourable for adsorption to the hydrophobic surface of the gold particle. Protein adsorption to the gold particles therefore results from a combination of hydrophobic binding and van der Waals forces.

In the examples described here, protein–gold conjugates were prepared with fibrinogen for ligand–gold labelling, and with whole IgG molecules or with Fab fragments of IgG for immunogold labelling. Fibrinogen was purified from fresh or frozen plasma by precipitation with 25% saturated ammonium sulphate followed by DEAE–cellulose chromatography (22). Anti-human fibrinogen, polyclonal IgG, was purchased from Sigma Chemical Co. (St Louis, Missouri) or from Cooper Biomedical (Malvern, Pennsylvania). Monoclonal antibodies specific for class I and class II antigens on macrophages were purified from the culture supernatants of the 34-1-2 and Mk-D6 hybridoma cell lines (American type culture collection), respectively, and purified by Protein A affinity chromatography. Fab fragments of IgG were prepared by papain (Immobilized papain, Pierce, Rockford, Illinois) digestion of polyclonal sheep anti-human fibrinogen (ICN Immunobiologicals, Lisle, Illinois). IgG was digested at 37 °C for 5 h in phosphate buffer containing 10 mM EDTA and 20 mM cysteine HCl at pH 7.0. Digested IgG was separated from the immobilized papain by centrifugation. Fab fragments were separated from Fc fragments by passage through an immobilized protein

G column (Pierce, Rockford, Illinois). The pH of gold solutions to be used in the preparation of protein–gold conjugates was adjusted with 0.2 M K_2CO_3 to pH 6.5 for fibrinogen, 7.3 for whole antibodies, and 6.5 for Fab fragments of antibody. In practice, the optimal pH for conjugation of a protein is determined empirically, by preparing a pH-dependent adsorption isotherm. With most proteins, optimal adsorption occurs slightly basic to the isoelectric point, although exceptions occur and the shape of the isotherm varies between protein species. Below the isoelectric point, the protein carries a net positive charge and will cause spontaneous flocculation of the negatively charged gold. This produces a colour change from red to blue, due to the increase in effective particle size as the particles form aggregates. Above the isoelectric point the protein carries a net negative charge, which increases in strength as the pH increases. This results in a charge repulsion which can prevent the adsorption of the protein to the gold at sufficiently high pH values. Addition of a small amount of electrolyte causes flocculation of unstabilized gold particles, providing an indicator of inadequate adsorption. The degree of stabilization can be evaluated visually or, preferably, photometrically, the colour changing progressively from red to blue as the degree of flocculation increases.

For spectrophotometric evaluation of the degree of stabilization, the absorbance is measured at 580 nm, where the absorbance of the flocculated gold reaches a maximum. Stabilization is indicated by an absorbance reading approaching zero. The amount of protein necessary to stabilize the gold is similarly determined by adsorption isotherms. A series of protein concentrations ranging from 1 to 100 µg are mixed with 1 ml aliquots of gold and a few drops of 10% NaCl solution are added. Di- and tri-valent ions are also effective at considerably lower concentrations. Where the amount of protein added is insufficient to completely coat the gold particles, the particles flocculate upon addition of the salt. Because the London–van der Waals forces acting on the very small gold particles are much weaker, flocculation of small particles occurs much more slowly. The lowest protein concentration which prevents flocculation of the gold in the presence of the NaCl is used for conjugation. In the situation where a small gold particle is to be conjugated to a somewhat larger protein molecule, as in 5 nm fibrogen–gold or 3 nm IgG–gold, conjugation using the isotherm concentration of protein produces a conjugate consisting of a mixture of one to several particles per molecule. In the preparation of 5 nm fibrinogen–gold conjugates, a three fold excess over the isotherm concentration of protein is used to prepare conjugates having primarily a single gold particle bound to each fibrinogen molecule, and a three fold excess of gold is used to prepare conjugates with several gold particles bound to each molecule. Secondary stabilizing agents such as bovine serum albumin or polyethylene glycol (20 000 mol. wt. linear dimer) are frequently added to the conjugates. These small molecules are thought to fill in the spaces between larger molecules on the gold particle surface, thereby further stabilizing the conjugate. These help prevent flocculation of the gold particles, but may not be necessary if the protein molecules are small enough to completely cover the gold surface, or when the gold particle is much smaller than the protein molecule, in which case the particle becomes embedded in the molecule.

Protocol 4. Determination of the optimal pH for conjugation of protein to colloidal gold

1. Adjust pH of 1 ml aliquots of the gold colloid to a series of values ranging from slightly below the isoelectric point of the protein to somewhat above it. The pH should be measured with a gel-filled pH electrode or with litmus paper, as the particles will flocculate on and plug a liquid-filled electrode. Acetic acid or K_2CO_3 can be added to adjust the pH.

2. Dilute sufficient protein to stabilize 1 ml of gold to 100 μl in each of a series of tubes. Ten to 20 μg of IgG will typically be sufficient to stabilize 1 ml of 18 or 30 nm gold, two to three times this amount will be required to stabilize 5 nm particles.

3. Add 1 ml of gold at each pH value with mixing to the series of tubes and allow the tubes to stand.

4. After 15 min add 0.1 ml of 10% NaCl and note the colour change. A shift in colour toward blue is indicative of inadequate adsorption of the protein to the gold. Below the isoelectric point of the protein, there may be a spontaneous colour change prior to the addition of the salt, and the mixture may take on a cloudy appearance. Most proteins will stabilize the gold at pH values at or above their isoelectric points, but some will be found to stabilize the gold over only a narrow pH range.

5. Choose the pH at which there is no colour change for preparation of the protein–gold conjugates.

Protocol 5. Determination of the minimal protein concentration needed to stabilize a gold sol

1. Filter the protein through a microporous filter or centrifuge at 100 000*g* for 30 to 60 min to remove debris and protein aggregates.

2. Prepare a series of protein concentrations in 100 μl of distilled water or dilute buffer. Excess electrolyte will cause flocculation of the gold, but up to 10 mM sodium phosphate (or lower concentrations of buffers containing di-valent or tri-valent cations) can be used if necessary to stabilize the protein.

3. Add 1 ml of gold of appropriate pH to each tube and gently mix.

Protocol 5. *Continued*

4. After a few minutes add 0.1 ml of 10% NaCl. Gold with no protein will quickly turn blue as the particles flocculate. As the protein concentration increases, the colour change will be less pronounced. The minimum concentration of protein which prevents flocculation, as measured by the colour change on addition of the salt, should be used for conjugation.

While larger gold sols will quickly turn blue on addition of excess electolyte, smaller (3 to 5 nm) particles are more stable in the presence of electrolyte and more time must be allowed for the colour change to occur before the isotherm can be evaluated. Gold sols prepared by the tannic acid method are particularly stable, and 3 nm tannic acid gold will not turn colour, but will merely become cloudy as the particles flocculate. Isotherms prepared with 18 nm gold sols can reasonably be evaluated within 15 min of the addition of the 10% NaCl, but isotherms with small gold particles should be allowed to stand for an hour at least, and preferably overnight.

Protocol 6. Preparation of the conjugate

When the appropriate pH and protein concentration for conjugation have been determined, the final conjugate can be prepared.

1. Dilute the protein such that 100 μl of protein solution will stabilize 1 ml of gold sol. This ratio ensures adequate mixing when the two are combined.

2. Dialyse the protein to reduce the electrolyte concentration of the solution. Dialysis for an hour against distilled water is preferable. If the protein in question will not tolerate this, addition of 5 mM sodium phosphate or Tris (tris[hydroxymethyl]aminomethane) buffer will usually stabilize the protein.

3. Centrifuge the protein for 30 to 60 min at 100 000*g*, or filter it through a microporous filter to remove debris and protein aggregates. Some adsorption to the filter should be expected, however.

4. Filter the required amount of gold sol through a microporous filter.

5. Place a suitable quantity of protein in a clean centrifuge tube together with the corresponding quantity of gold sol and gently mix.

6. Allow the mixture to stand for 15 to 30 min.

7. Add secondary stabilizing agents such as polyethylene glycol or bovine serum albumin after filtration through a microporous filter. Albumin is prepared as a 10% solution in water and added to give a final concentration of 1%. If polyethylene glycol is to be used as the secondary stabilizer, 50 μl of a 1% solution is added for each ml of gold sol.

8. Centrifuge the conjugate to concentrate it and to separate gold labelled protein from unlabelled protein. Three and 5 nm gold particles are pelleted by centrifugation at 50 000*g* for 1 h. Eighteen and 30 nm gold particles are spun at 10 000*g* for 30 min. The conjugate should form a loose pool rather than a firm pellet, and any aggregated material should be discarded.

9. Discard the supernatant and re-suspend the conjugate to approximately 20% of the original volume in a buffer suitable to the protein and the use to which it will be put. While the protein coating on the gold inhibits flocculation due to the presence of electrolytes, the conjugate will remain stable for longer periods in lower molarity buffers.

3. Imaging colloidal gold by light microscopy

3.1 The colloidal gold markers

As first shown by Allen (23) and Inoue (24, 25), detectability thresholds for low contrast objects in the light microscope can be considerably reduced by the application of video techniques for contrast enhancement. This procedure is particularly well suited for application to the differential interference contrast (DIC) image. While colloidal gold particles of sizes typically used for cell surface labelling are an order of magnitude smaller than the Abbé limit and thus cannot be resolved by the light microscope, they produce an inflated diffraction image which can be detected by the use of a high resolution video camera and subsequent contrast enhancement of the video signal. Individual labels as small as 10 nm in size can be detected. Groups of smaller labels can be followed as they impart a darkening to the cell surface, as they are massed inside a vesicle following internalization, or are concentrated at other specific locations on or in the cell.

Small gold labels can also be silver enhanced to improve their visibility in the light or electron microscope (references 26–28 and Chapters 3, 4). This method uses the colloidal gold particle as a nucleus to create a silver particle of larger size in a process similar to photographic development. In general, silver enhancement requires a source of silver ions and a reducing agent. The reaction is generally performed at an acidic pH and in the presence of a protective colloid to control the rate of the reaction and uniformity of the reaction product. Silver lactate is commonly the source of silver ions although silver acetate, silver nitrate, and certain film emulsions have also been used (29–31). Hydroquinone is primarily used as the reducing agent, though the employment of other agents including bromohydroquinone, ascorbic acid, and 4-methylaminophenolsulphate has also been reported. Citrate, lactate, or acetate buffers all appear equally capable of producing the acidic conditions required for the silver enhancement reaction to proceed.

This technique is useful when, due to steric considerations, small labels are required to achieve adequate labelling, or as a quick screening method for examination by light microscopy to determine the success of a labelling protocol for electron microscopy. There are, however, several disadvantages to silver enhancement. The method cannot be used to image gold particles on living cells because the solutions involved are acidic and specimens must be fixed prior to silver enhancement. Copper grids will interfere with the reaction in some protocols, necessitating the use of gold or nickel grids. In addition, even though particle size can be controlled such that the silver particle grows large enough for easier detection in electron microscopes or even for resolution of single particles by light microscopy, the size of the silver reaction product limits the spatial resolution of the label for electron microscopy. Also, particles may grow extremely rapidly or unevenly, adjacent particles may fuse as the silver granules grow, and particles may be enhanced selectively. These may not be serious problems at the light microscopic level, but they can limit the usefulness of this technique for high-resolution or quantitative studies. In this case the addition of one of several protective colloids will enable the reaction to proceed in a controlled fashion resulting in a slower growth rate and a more uniform particle size (29, 30, 32, 33). Gum arabic, also known as gum acacia, a complex co-polymer of arabinose, galactose, rhamnose, and glucuronic acid, is generally preferred (*Protocol 11*, Chapter 3). Alternatively, polyethylene glycol, polyvinyl pyrrolidone, gelatin, dextran, bovine serum albumin, and gum tragacanth have been used. Without protective colloid, enhancement is non-linear, irregular, and can produce a 5000 to 10 000 fold size increase within 10 min. With the addition of colloid a more even 10 fold enhancement can be achieved in 20 to 30 min.

A number of procedures have been described for silver enhancement. The Danscher (32) methodology and modifications of it are perhaps the most widely used. The Ilford L-4 (silver halide) film emulsion with a Metol (phenosulphate) or thiocyanate developer has proved useful for enhancement of smaller bead sizes employed in high resolution localization work (31). The L-4 procedure is reported to produce a uniform two to three fold increase in particle diameter in 10 to 15 min. A number of commercial kits are also available. Stierhof (28) compared enhancement of 1.4 nm and 4 nm gold particles by the Danscher and Ilford methods with that obtained with a commercially available kit, and found that, while some degree of irregularity of enhancement was seen with each technique, the original Danscher method gave the most even particle growth. Enhancement with the Ilford technique resulted in slightly greater variability, particularly in areas of high particle density. The commercial kit produced very rapid and variable particle growth, though this could be overcome by the addition of gum arabic. While the first two methods are light sensitive and must be performed in a darkened room illuminated with a red safelight, the commercial kit is not light sensitive and thus permits monitoring of the progress of enhancement with a light microscope. It should also be noted that post-fixation with OsO_4 will reverse the enhancement process, decreasing the size of the silver particles. For a description of the silver enhancement technique refer to *Protocol 11*, Chapter 3 and *Protocol 8*, Chapter 4. For each particular application the degree of enhancement will vary with the initial gold particle size and the time required to achieve the desired final particle size should be determined empirically.

3.2 Design of the light microscope for video-enhanced differential interference contrast observations

An example of a light microscope based on the designs of Inoue (24, 25) and currently in use in our laboratory is shown in *Figure 1*. The inverted microscope was built on a Nikon Diaphot (Garden City, New Jersey) modified for increased light transmittance to the video camera. This was achieved by replacing the standard image splitting prism, which normally directs 80% of the light to the camera and 20% to the oculars, with the front surface mirror which directs 100% of the available light to the video camera. The microscope is equipped with a 1.4 numerical aperture polarization rectified condenser (25) and DIC optics (23, 24, 34). Illumination is supplied by a 100 W high pressure mercury source used with a 546 nm interference filter and a diffuser. When DIC is used, a small degree of bias retardation with the Wollaston prism adjusted, the increased image brightness and hence reduced image contrast is helpful for imaging gold particles. It is also possible to observe

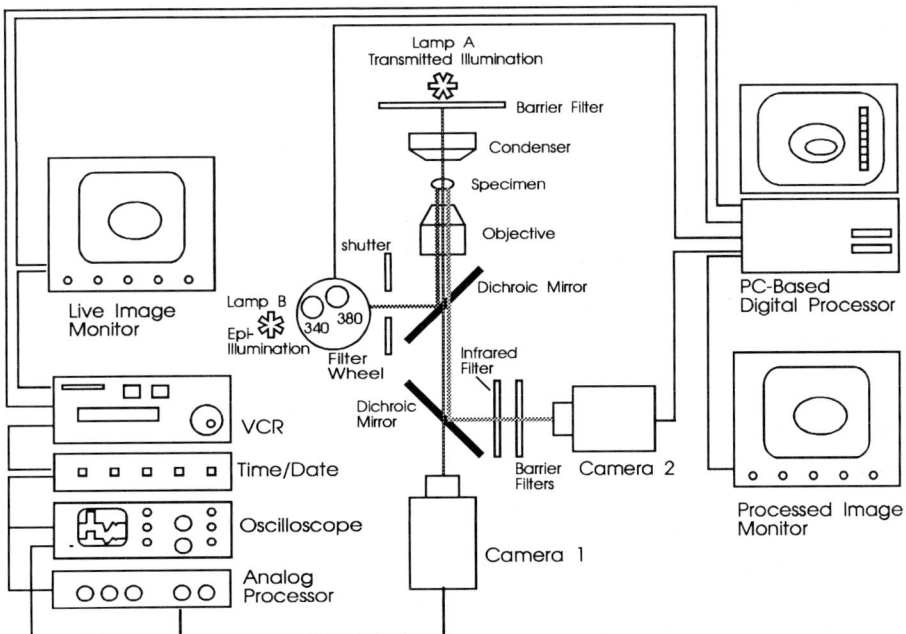

Figure 1. Schematic drawing of video-enhanced light microscope.

colloidal gold-labelled cells with bright field illumination and with asymmetric illumination contrast imaging (35). Dual wavelength imaging with appropriate dichroic and barrier filters is also possible for studies utilizing fluorescent probes for ionic or pH changes combined with labelling of structures.

The video signal is collected with an instrumentation grade Newvicon® (Matsushita Electric Co. camera tube with photoconductive target composed of cadmium and zinc tellurides) video camera modified for manual control of gain and black level (Dage-MTI, Michigan City, Indiana). Manual control is necessary because the automatic controls will continuously adjust gain and black level to compensate for changes in the image (Chapter 7, Section 3.2.3) making it impossible to compare brightness between video frames to detect changes that occur as a consequence of the binding of colloidal gold to the cell surface. An analogue video processor provides non-linear expansion of contrast and brightness in the video image. Images are recorded for later playback and analysis on a high-resolution 3/4" U-matic and on a time-lapse video recorder which facilitates detection of movement and processes which are difficult to discern by observation of the experiment in real time. Digital processing of the video allows further contrast enhancement as well as background subtraction, shading correction, and real time image averaging to reduce noise in the image. Images can be digitized and processed from either live or recorded video. Image processing and analysis can be performed on PC-based systems with high processing speeds and large, fast storage capabilities with packages such as Image-1 (Imaging Technology, Woburn, Massachussetts) which include an image processing board and software.

To facilitate subsequent electron microscopic examination of the same cell which was observed by VDIC, cells are allowed to adhere to finder-patterned, filmed, TEM grids. Finder grids, available from several suppliers, are produced with a marked grid pattern permitting the accurate location of the cells. Nickel grids prove sufficiently non-toxic for culturing most cells, though gold and polymer grids are also available. Typically, Formvar (polyvinyl formal)-coated grids are used, though thin films of other polymers such as polystyrene or various polyurethanes can be cast and often will provide improved cell adhesion (36, 37). The grid is placed on a slide (or, for the inverted microscope, a coverglass) on the light microscope stage and held in place by strips of Parafilm or double-sided sticky tape which just overlap the edges of the grid. By placing narrow strips cut from a coverglass on each side on the grid as spacers under the top coverglass, a small flow chamber is created, permitting the addition of gold label and of other solutions for washing and fixing cells while they are being observed.

4. Imaging colloidal gold by electron microscopy

Due to its electron density and regular size and shape, colloidal gold is readily identified by either scanning or transmission electron microscopy. To show the final positions of gold labels relative to internal features of the cell (*Figures 2d* and *3d*) whole mounts of cells can be examined with a high-voltage electron microscope at 800–1200 kV. Stereo pair images taken at 7 ° tilt angle reveal the three dimensional architecture of cytoskeletal elements and internal cell ultrastructure (42).

Images of cell surfaces are obtained by examination of the cells or tissues with a scanning electron microscope. In the examples described here, whole mounts of cells were examined with a Hitachi S-900 field emission scanning electron microscope (LVSEM). The field emission gun produces an electron beam of small size and of sufficient brightness to permit high resolution imaging at low beam voltages (13, 38, 56). Observation of specimens at accelerating voltage (V_0) of 1 to 2 kV, where beam penetration is on the order of a few nanometres, provides an image of the membrane surface and of the surface of the protein coating of gold labels bound to the cell (13, 39). Gold labels can be identified by their shape, but cannot be distinguished from other surface features of similar size and shape. Increasing V_0 to 3–5 kV and above increases beam penetration, progressively decreasing the amount of surface detail in the image but making possible positive identification of gold labels through atomic number (Z) contrast (12, 13, 40). Heavy metal-stained sub-surface structural elements also become visible as V_0 is increased. The use of backscattered electrons (BSE) for imaging further increases differences in Z contrast. With conventional SEMs, accelerating voltages of 20 to 30 kV are required for BSE imaging of particles of 10 to 20 nm diameter, yielding an image in which gold particles are easily identified but containing very little information on the structure of the cell surface. At low kV values the ratio of primary to backscattered electrons is improved but, because of the low energy of the BSE, they are difficult to detect with standard BSE detectors. Backscattered electron detectors using the YAG-type scintillator provide much higher detection efficiency. Recent improvements involving the use of conductive, reflective, and diffusive coatings on the surfaces of these scintillators to prevent surface charging and to increase light transmittance to the photomultiplier tube have made possible the production of BSE detectors of higher sensitivity such that BSE images of gold particles can be obtained at voltages down to 2.5 kV (47, 54, 55). BSE imaging of gold-labelled cell surfaces at lower beam voltages permits positive identification of gold labels down to 1 nm in size and a concomitant increase in the amount of surface detail in the image as the voltage is decreased.

Protocol 7. Preparation of specimens for electron microscopy

1. Fix specimens for 30 min in 0.1 M Hepes-buffered 1% glutaraldehyde containing 0.5% tannic acid for stabilization of cytoskeletal elements. For improved penetration of tannic acid into the cells, 0.05% saponin can be included in the fixative to permeabilize cell surface membranes, though

Protocol 7. *Continued*

this sometimes creates openings in the membrane large enough to be visible in the high resolution scanning electron microsope.

2. Rinse in several changes of buffer, and post-fix the cells in Hepes-buffered 0.05% OsO_4 for 15 min.

3. Rinse the specimen in distilled water and stain with 1% aqueous uranyl acetate for an additional 15 min. (Specimens intended only for observation by scanning electron microscopy may be fixed in buffered 1% glutaraldehyde and subsequent treatments with OsO_4 and uranyl acetate omitted, otherwise preparation is identical for specimens intended for either type of microscopy.)

4. Dehydrate specimens through a graded ethanol series, typically 30%, 50%, 70%, 80%, 85%, 90%, 95% for 3 to 5 min at each step, ending in several changes of 100% ethanol which has beed dried by storage over molecular sieve (Type 3A, Linde Division Union Carbide, Danbury, Connecticut).

5. Dry the specimens by the critical point method using CO_2 as the transitional fluid (41, 42). The critical point dryer should be equipped with an in-line molecular sieve trap (for example #8782 Tousimis Research Corp, Rockville, Maryland) and a hydrophobic filter (Type AP25 Millipore) to exclude water from the chamber. Alternatively, specimens can be prepared by ultra-rapid freezing, at rates sufficiently high to minimize damage from ice crystal growth, and freeze drying (43, 44). Care must be taken to ensure that complete dehydration occurs below temperatures known to permit collapse of partially dehydrated material. Store dried samples over molecular sieve and/or *in vacuo* (43).

6. Observe specimens by LVSEM uncoated or coated with a thin layer of platinum to improve the signal arising from the cell surface. Apply 10 to 20 Å of platinum by ion beam sputtering (45, 46) with tumbling and rotation of the specimen during coating or by planar magnetron sputtering (47). This thin coating permits high resolution imaging of cell surface detail and also allows specimens to be examined by HVEM after metal coating (48). Alternative imaging protocols employing higher accelerating voltages in combination with chromium coating have also been reported (49, 50).

---

5. Labelling with colloidal gold

To demonstrate the use of colloidal gold labelling in conjunction with VDIC, HVEM, and LVSEM, two model systems were employed. In the first of these, platelets obtained from normal healthy adult volunteers were introduced to a flow cell containing a Formvar-coated Ni Maxtaform finder grid on the light microscope stage. The degree of activation and spreading was monitored as platelets settled and adhered to the grid (*Figure 2*). Excess platelets were rinsed away with Tyrodes buffer before gold label was introduced to the chamber. Surface-activated, spread platelets were labelled with fibrinogen or fibrinogen–gold conjugates (FgnAu) while under observation in the VDIC. The protein or conjugate was applied for 5 min, unbound label was rinsed away with Tyrodes buffer, and the platelets were incubated in buffer for an additional 5 min to allow bound receptor/ligand complexes to complete their movement across the platelet surface before fixation with glutaraldehyde and preparation for electron microscopy. To identify soluble fibrinogen bound to the platelet surface, fibrinogen-labelled platelets were fixed with 0.05% glutaraldehyde in Hepes buffer; free aldehyde groups were quenched with 0.05 M glycine in Hepes buffer for 5 min and the fibrinogen on the platelet surface was labelled with anti-fibrinogen conjugated to 18 nm gold particles or with Fab fragments of anti-fibrinogen conjugated to 3 nm gold particles (FabAu$_3$) (*Figure 3*). Control platelets not exposed to exogenous soluble fibrinogen were labelled only with 18 nm anti-fibrinogen gold or FabAu$_3$.

For the second model system, class I and class II histocompatibility antigens on macrophage surfaces were immunolabelled with colloidal gold (*Figure 4*). Inflammatory peritoneal macrophages were harvested from adult Balb/c mice and cultured on Formvar-coated finder grids. Ia (class II) antigen expression was induced by incubating the macrophages for 96 h in DMEM + 3% supernatant from concanavalin A-activated rat spleen cells (Con A sup.).

Grids bearing cultured macrophages were placed in a drop of Dulbecco's medium on a coverglass, shims made from strips cut from a coverglass were placed on each side of the grid, and another coverglass was placed on top of the shims, created a flow cell with the grid inside. Antibody to class I antigen conjugated to 12 nm colloidal gold and antibody to class II antigen conjugated to 24 nm colloidal gold were added to the flow cell separately or together. Due to the complex structure of the cells, individual labels could be followed only near the cell margin where the cells were thinner and relatively smooth. Most of the 12 nm class I labels remained on the cell surface and were transported to large patches over the body of the cell, appearing as shadowed areas developing over the course of several minutes. The 24 nm class II labels were rapidly internalized, filling vesicles which were subsequently transported to the cell centre surrounding the nucleus. The cells were fixed with 1% glutaraldehyde at various intervals following the addition of label and prepared for electron microscopy. The principle of labelling with gold probes in these two examples is identical to that described in Chapters 3 and 4.

Labelling efficiency is a function of label concentration, binding coefficient, viscosity, temperature, and physical mixing (51, 52). However, when live cells are being labelled the only parameter that can be altered to quickly and efficiently label the cells is the concentration of label. Hence, to label the

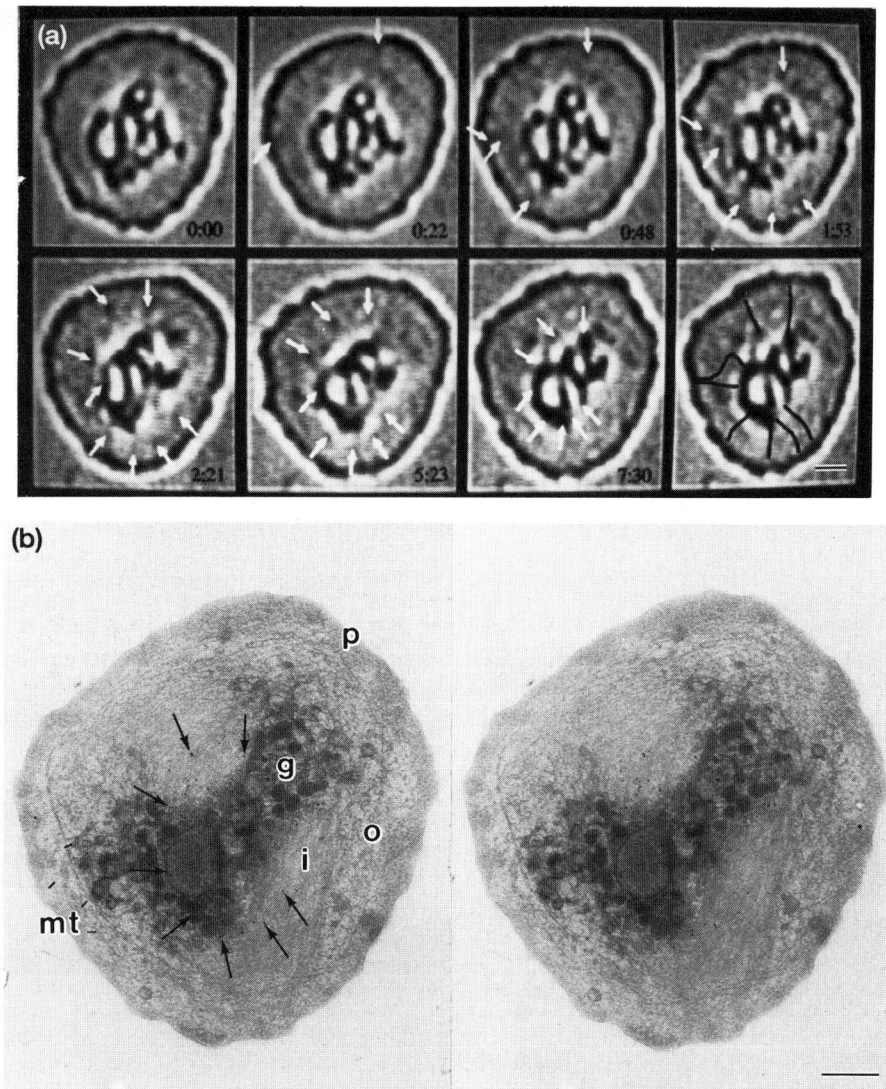

Figure 2. Platelet labelled with a dilute suspension of 18 nm fibrinogen–gold. (a) Video-enhanced light micrograph series. Gold labels are detectable by their inflated diffraction image as they bind in a diffuse pattern on the platelet surface and trigger centripetal movement of the fibrinogen/receptor complex. Arrows indicate selected particles binding near the platelet margin and moving across the platelet surface toward the granulomere, seen as a dark area in the centre of the platelet. Times are elapsed times after addition of gold label. Unbound gold was rinsed away after 3 min of exposure. Final frame shows paths followed by the indicated particles. Movement is directed, generally following a more or less straight line, though occasional gold labels will pause and drift sideways when they cross the line between the inner and outer filamentous zones of the subjacent cytoskeleton. Scale bar 1 μm. (b) 1 MeV transmission electron micrographic

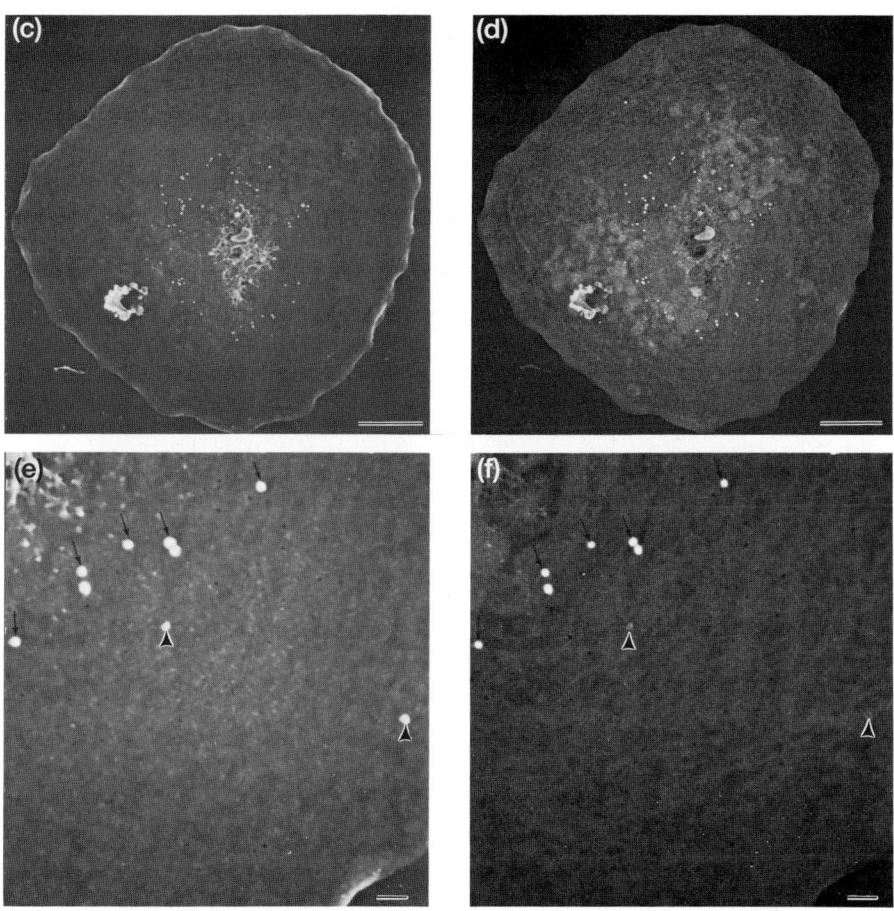

stereo pair of same platelet shown in (a). Platelet granules and concentric zones of the cytoskeleton are readily distinguished. Peripheral web (p), is a non-substrate-attached, dense band of short actin filaments at the platelet margin. Outer filamentous zone (o) is comprised of loosely woven actin filaments and the inner filamentous zone (i) is a more densely woven zone of actin filaments surrounding the granulomere (g). Occasional microtubules (mt) can be seen running through the outer and inner filamentous zones. Gold labels appear as small black spheres overlying the inner filamentous zone. Arrows indicate labels followed in VDIC. Scale bar 1 μm. (c) Same platelet imaged at 1.5 kV accelerating voltage in the field emission SEM. Beam penetration at this low voltage is slight, and the signal arises primarily from the membrane surface and the protein coating the gold particles, providing confirmation that the gold labels remain on the platelet surface. Scale bar 1 μm. (d) Imaging the platelet at 4.5 kV in the backscattered electron imaging mode permits positive identification of the gold labels. Increased beam penetration leads to signal arising from internal structures of platelet. Scale bar 1 μm. (e) Higher magnification of area of platelet at 1.5 kV. Gold particles (arrows) are identified by shape alone and cannot be distinguished from other structures of similar size and shape (arrowheads). Scale bar 0.1 μm. (f) Same area as (e) at 4.5 kV. Atomic number contrast now permits distinction of the dense gold particles (arrows) from similarly sized structures (arrowheads). Scale bar 0.1 μm.

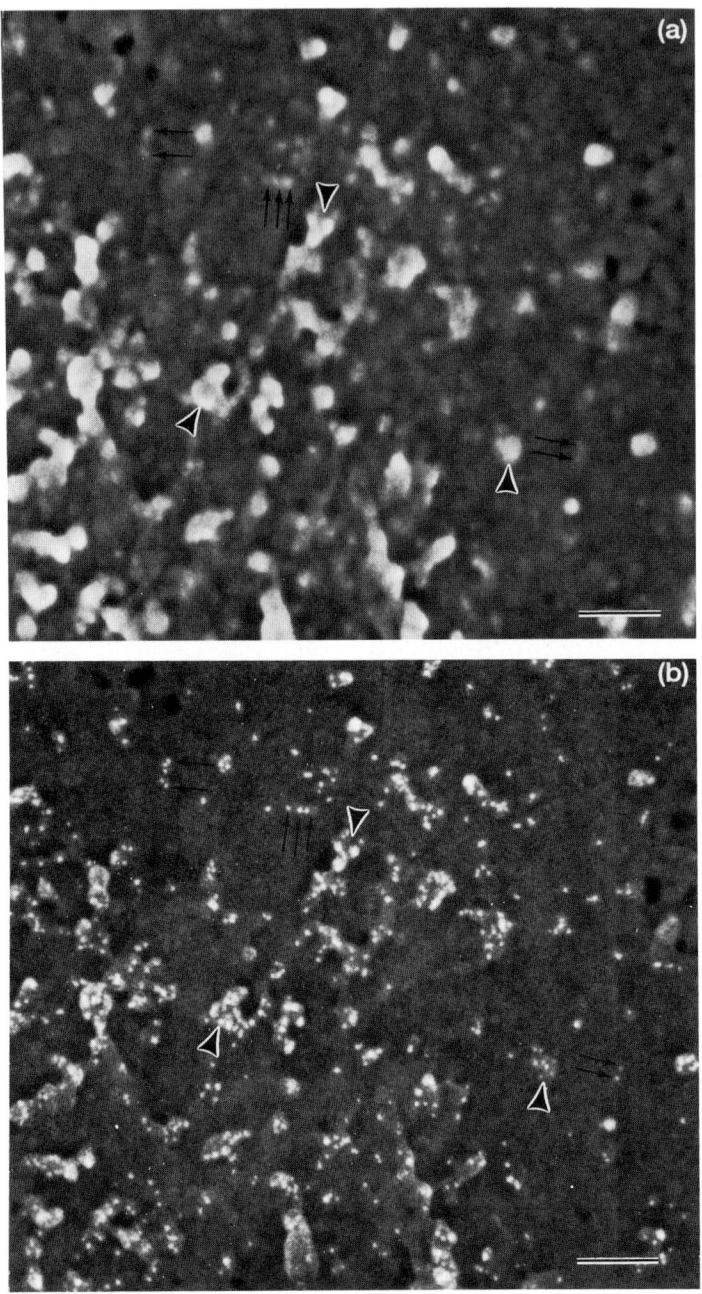

Figure 3. Fab fragments of anti-fibrinogen IgG conjugated to 3 nm gold particles, labelling fibrinogen bound to platelet surface receptors. (a) SEM at 1.5 kV. Single fibrinogen molecules (arrows) as well as aggregates of protein (arrowheads) can be seen on the platelet surface. (b) Backscattered image at 4.5 kV. The gold-conjugated Fab fragments outline the molecules (arrows) and aggregates (arrowheads) of fibrinogen. Scale bar 0.1 μm.

greatest number of receptors simultaneously, the ligand-gold must be added in high concentration. For these experiments gold label is typically added at a concentration of approximately 2×10^{13} particles/ml, so that maximal labelling efficiency occurs within 2 to 5 min. For conventional labelling of fixed material maximal labelling can be attained at lower (5×10^{12}) particle concentrations and 20 to 30 min incubation times. Still lower concentrations required incubation periods of hours to days to reach maximal labelling efficiency. Horisberger and Rosset (53) provide a table for relating the concentrations of colloidal gold particles of different sizes with their optical absorbances.

Single FgnAu labels of 12 to 30 nm diameter could be followed by VDIC as they bound in a dispersed pattern to the platelet surface, then moved in the plane of the membrane toward the granulomere. Individual 5 nm labels were too small to detect by VDIC, but groups of 5 nm FgnAu labels could be seen moving across the platelet surface, and the concentration of small gold near the platelet centre could be seen by the darkening of the image as the gold-labelled receptors collected in a band over the inner filamentous zone surrounding the granulomere. When platelets were labelled with soluble fibrinogen followed by anti-fibrinogen conjugated to 18 nm gold labels, the gold was seen to bind directly to the area over the inner filamentous zone, demonstrating that the soluble fibrinogen had followed the same pattern of binding and centralization as the fibrinogen–gold conjugates. Labelled platelets were fixed with glutaraldehyde while still being observed and the position of the platelet on the finder grid noted. The specimen was prepared for electron microscopy.

As platelets attach and spread on a surface, their cytoskeletons undergo a reorganization into a series of concentric zones (14). At the margin of the spread platelet is a dense meshwork of short actin filaments which has been termed the peripheral web. Inside the peripheral web are two circumferential zones of longer actin filaments and microtubules. The outer filamentous zone is a loosely organized network of actin filaments with occasional microtubules running through it, and the inner filamentous zone is a more tightly woven meshwork which surrounds the granulomere, a cluster of platelet granules which may be centred or off to one side of the spread platelet. Examination of FgnAu-labelled platelets with the HVEM showed the fibrinogen/receptor complexes to be collected in a band overlying the inner filamentous zone of the cytoskeleton and surrounding the granulomere. Subsequent examination by LVSEM of the same platelet confirmed that the gold was situated on the outer surface membrane. Operation of the instrument at low, 1–1.5 kV, beam voltages provides detailed images of the membrane surface and the surface of the protein coat on the gold particles. The particles could be identified by shape alone but could not be distinguished from other structures of similar size and shape on the platelet surface. Higher accelerating voltages, 2.5 kV and above, produce greater beam penetration into the specimen and

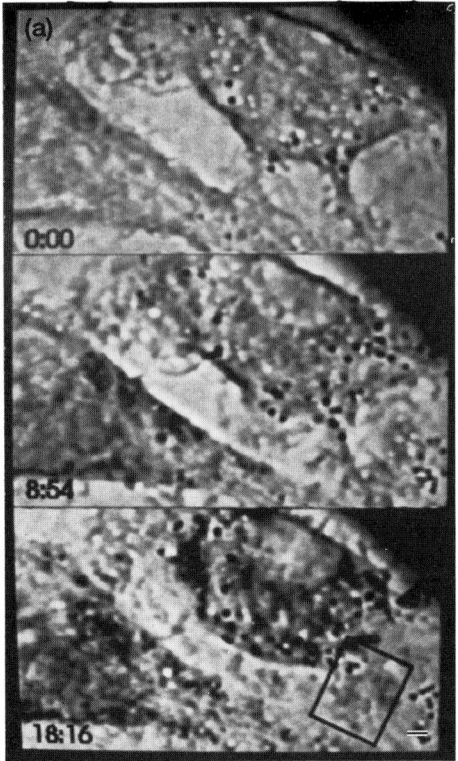

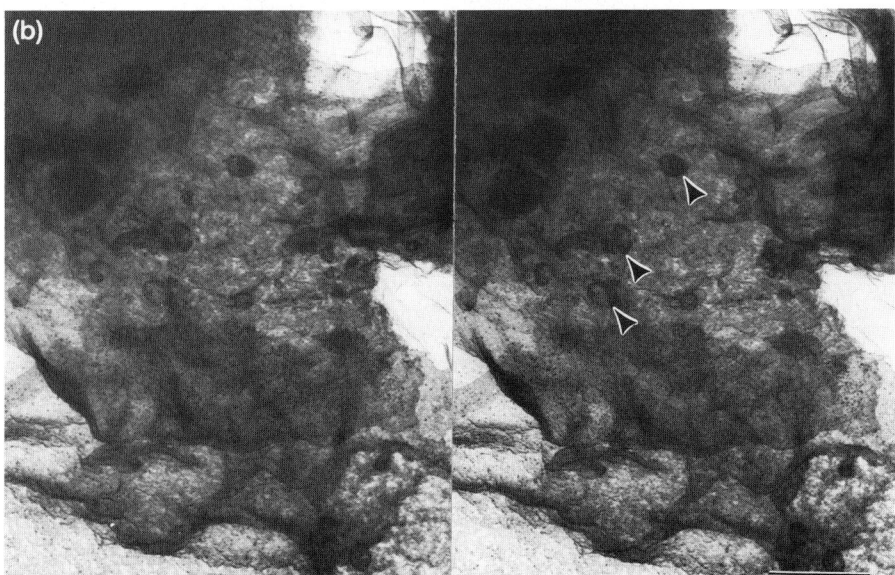

Figure 4. Mouse peritoneal macrophage labelled with 12 nm, colloidal gold conjugated to antibody to the class I MHC antigen. (a) VDIC series. Times are elapsed times after the

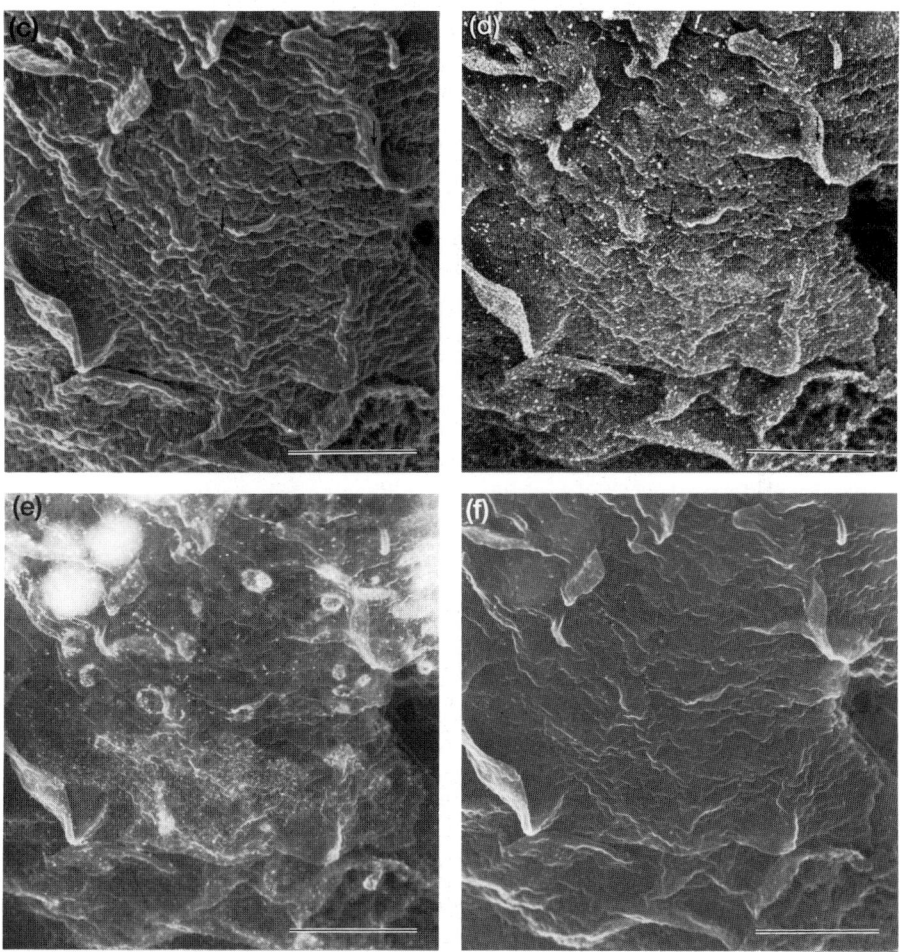

addition of gold label. Individual small particles are not readily ditinguished on the surfaces of these large, complex cells. Much of the class I antigen remains on the cell surface after labelling, forming large patches seen as shadowed areas. Vesicles filled with internalized gold label can be seen to surround the nucleus. The box on the lower panel indicates the area shown in subsequent electron micrographs. (b) HVEM stereo pair of area indicated in box in (a). Gold label appears as small black spheres on the cell surface, lining the interiors of vesicles (arrowheads), and forming a densely labelled patch between two cells (c). Low (1.5 kV) accelerating voltage (V_0) SEM of same area shown in (b). Gold labels (arrows) appear as small white spheres on the cell surface. (d) Increasing V_0 to 4.5 kV in the backscattered electron imaging mode permits easy detection and positive identification of the dense gold particles (arrows). Beam penetration is increased, but most of the signal arises near the surface. (e) Increasing V_0 to 20 kV in the backscattered mode increases beam penetration and permits imaging of sub-surface gold and heavy metal-stained cell structures. (f) Imaging the cell surface at 20 kV in the secondary electron imaging mode yields less surface detail than low kV secondary images and less atomic number contrast than backscattered images. Scale bar 1 μm.

171

were used for positive identification of the dense gold particles by atomic number contrast.

Gold-labelled macrophages were similarly examined with the HVEM followed by platinum coating and examination by LVSEM. Observation by HVEM revealed both 12 nm class I and 24 nm class II label internalized in vesicles, though most of the class I label remained on the cell surface. When both label types were added simultaneously, the class I and class II antigens were found to segregate into separate vesicles during internalization. Detection by LVSEM of gold labels on the rough surface of the macrophage is more difficult than on the relatively smooth surface of the platelet, particularly when the instrument is operated at low beam voltage and labels are identified primarily by shape. In this situation the use of higher beam voltages and especially of the backscattered electron detector is essential for positive identification of gold labels and to ensure that a high proportion of the labels are detected.

With the LVSEM individual protein molecules can be imaged directly on the platelet surface. To confirm that the gold-conjugated fibrinogen was retaining its activity, that the binding and movement seen with fibrinogen–gold was representative of a receptor/ligand interaction rather than a result of the fibrinogen being bound to the gold particle, soluble fibrinogen was used to label the fibrinogen receptor on spread platelets. When the soluble fibrinogen was given time to bind and be centralized on the platelet surface, small protein aggregates were seen on the surface over the inner filamentous zone, in a similar pattern to that seen with FgnAu. To monitor the formation of these aggregates, platelets were fixed at various time intervals after the addition of the fibrinogen. At 1–1.5 min, single rod-shaped fibrinogen molecules could be seen scattered on the platelet surface. After 3 min of incubation, small clusters and aggregates had begun to form on the surface and move toward the granulomere. These aggregates continued to grow and move across the surface until all of the receptors were bound and included in the aggregates collected in a band over the inner filamentous zone surrounding the granulomere (57). Immunogold labels were used to confirm the identity of the fibrinogen on the platelet surface. Large labels, 18 nm gold conjugated to whole IgG anti-fibrinogen, labelled the platelet-bound fibrinogen in clusters over the inner filamentous zone. These gold probes however, were nearly as large as the protein aggregates to be labelled. For better resolution, small gold probes were prepared by conjugating Fab fragments of anti-fibrinogen IgG to 3 nm gold spheres. The 3 nm particles are approximately half the size of the Fab fragments, and form a probe consisting of one Fab to one small gold particle. The Fab–gold conjugates were small enough to outline a single fibrinogen molecule on the platelet surface, several labels binding to one of the larger molecules. This resulted in both better resolution of the bound fibrinogen and in much higher labelling density (40).

6. Summary

Colloidal gold probes of virtually any size can be used for correlative light and electron microscopy. However, the optimal size to be used depends on the structures to be labelled, experimental conditions, and the available instrumentation.

With the approach outlined here, colloidal gold labels that are smaller than the resolution limit of the light microscope may be imaged on live cultured cells, using video-enhanced light microscopy. The movement of gold-labelled receptors may be tracked on the cell surface or even into the cell interior using optical sectioning. Gold labelling can also be used to determine the final positions of ligand-bound unlabelled receptors by labelling the ligands after movement has occurred.

The largest gold labels, 100 nm diameter and greater, are large enough to actually be localized with the light microscope, but they provide low labelling efficiency and poor EM spatial resolution. Typically, labels used for light microscopic studies are in the 10–30 nm size range. These labels are large enough that individual labels can be detected by video-enhanced light microscopy and can readily be visualized by electron microscopy. Larger labels are desirable when detectability of the gold label is important, when labelling thick or complex cells, or when there are few sites to be labelled. While labelling efficiency is generally considered to be greater with smaller gold probes, this effect is often merely a result of the concentration of particles in the suspension. Because the same concentration of $HAuCl_4$ is used in the preparation of all sizes of colloidal gold and the type and quantity of reducing agent varied to produce different sizes, the larger particles are necessarily less concentrated in the resulting sols. The concentration differential will follow the third power of the particle radius. Thus, doubling the particle size results in an eightfold decrease in concentration. This differential will persist through the preparation of protein–gold conjugates unless specific care is taken to account for it.

Smaller, 1.4–5 nm diameter, gold particles are most useful when structures to be labelled are small and close together, when the path to the protein of interest is obscured or tortuous, or when high spatial resolution is desired. Care must be taken, however, to consider the size of the conjugate as a whole, and not just that of the gold particle. While larger gold particles generally have room for a number of protein molecules to adsorb to their surfaces, 1–5 nm gold particles may be smaller than the proteins to be conjugated. When the protein is much larger than the gold, as, for instance, in the case of 3 nm particles conjugated to whole IgG molecules, several gold particles may bind to a single molecule, depending on the ratio of gold to protein used in conjugation. Conjugation of 5 nm particles to the 15 nm IgG molecules produces a one to one ratio of gold particles to protein molecules.

The size of the resulting probe in this instance depends principally on the size of the protein molecule, and will not decrease significantly through the use of 1.4 or 3 nm particles. Thus, the sizes of both gold and protein must be considered when preparing small gold probes.

Individual small gold probes cannot be detected by VDIC, though concentrations of small particles can be seen as a darkening of the cell surface or as a dark vesicle following internalization, and movement of groups of small particles may be followed. The resolution of transmission electron microscopes and of the LVSEM are such that the smallest gold particles can be imaged however the high magnification necessitates a correspondingly small field of view. Larger particles may be necessary for visualization by conventional SEMs.

With this correlative microscopic approach to the labelling of cell surface receptors with colloidal gold, the behaviour of living cells and the complex dynamic actions of the receptors are observed. By fixing the cells while they are being observed and preparing them for subsequent examination by scanning and transmission electron microscopy, the ultrastructural features underlying those processes can be determined. With this methodology, complex relationships between cell adhesion and spreading, cell surface receptor occupancy, cytoskeletal reorganization, and cell–cell interactions may be investigated.

References

1. Wetzel, B. and Albrecht, R. M. (1989). In *The Science of Biological Specimen Preparation (Scanning Microscopy Supplement 3)*, (ed. R. M. Albrecht and R. L. Ornberg), pp. 1–6. Scanning Microscopy International, Chicago.
2. Albrecht, R. M., Simmons, S. R., and Malecki, M. (1989). *Proc. EMSA*, **47**, 904.
3. Albrecht, R. M., Olorundare, O. E., Bielich, H. W., and Simmons, S. R. (1987). *Proc. EMSA*, **45**, 446.
4. Albrecht, R. M., Olorundare, O. E., and Simmons, S. R. (1988). In *Fibrinogen. Vol III. Biochemistry, Biological Functions, Gene Regulation and Expression*, (ed. M. Mosesson, D. Amrane, K. Siebenlist, and J. Diorio) pp. 211–14. Excerpta Medica, Elsevier, Amsterdam.
5. Goodman, S. L. and Albrecht, R. M. (1987). *Scanning Microsc.*, **1**, 727.
6. Olorundare, O. E., Goodman, S. L., and Albrecht, R. M. (1987). *Scanning Microsc.*, **1**, 735.
7. Goodman, S. L., Park, K., and Albrecht, R. M. (1991). *Coll. Gold Methods*, **3**, 369.
8. Bendayan, M. (982). *J. Histochem. Cytochem.*, **31**, 81.
9. Albrecht, R. M., Goodman, S. L., and Simmons, S. R. (1989). *Am. J. Anat.*, **185**, 1.
10. Horisberger, M. (1981). *Scanning Electron Microsc.*, **II**, 9.
11. Sheetz, M. P., Turney, S., Qian, H., and Elson, E. L. (1989). *Nature*, **340**, 284.

12. Albrecht, R. M. and Hodges, G. M. (1988). *Biotechnology and Bioapplications of Colloidal Gold*. Scanning Microscopy International, Chicago.
13. Pawley, J. B. and Albrecht, R. M. (1988). *Scanning Microsc.*, **10**, 184.
14. Loftus, J. C. and Albrecht, R. M. (1984). *J. Cell Biol.*, **99**, 822.
15. Pawley, J. B. (1984). *J. Microsc.*, **136**, 45.
16. Guagliardi, L. E., Paulnock, D. M., and Albrecht, R. M. (1987). *Scanning Microsc.*, **1**, 705.
17. Albrecht, R. M., Oliver, J. A., and Loftus, J. C. (1991). In *The Science of Biological Specimen Preparation for Microscopy and Microanalysis*, (ed. M. Müller, R. P. Becker, A. Boyde, and J. J. Wolosewick), pp. 185–93. Scanning Electron Microscopy, Chicago.
18. Overbeck, J. (1952). In *Colloidal Dispersions*, (ed. S. W. Goodwin) pp. 1–22. Henry Ling, Dorchester.
19. Slot, J. W. and Geuze, H. J. (1985). *Eur. J. Cell. Biol.*, **38**, 87.
20. Roth, J. (1983). In *Techniques in Immunocytochemistry*, Vol. 2, (ed. C. G. R. Bullock, and P. Petrusz), p. 217. Academic Press, New York.
21. Frens, G. (1973). *Nature Phys. Sci.*, **241**, 20.
22. Nagatani, T. and Saito, S. (1986). *Proc. XIth ICEM Meeting*, 2101–8.
23. Allen, R. D., Allen, N. S., and Travis, J. L. (1981). *Cell Motility*, **1**, 291.
24. Inoue, S. (1981). *J. Cell Biol*, **89**, 346.
25. Inoue, S. (1986). *Video Microscopy*. Plenum, New York.
26. Scopsi, L. (1989). In *Colloidal Gold: Principles, Methods, and Applications*, Vol. 1, (ed. M. A. Hayat), pp. 251–95. Academic Press, San Diego.
27. Hacker, G. W. (1989). In *Colloidal Gold: Principles, Methods, and Applications*, Vol. 1, (ed. M. A. Hayat), pp. 297–321. Academic Press, San Diego.
28. Stierhof, Y-D., Humbel, B. M., and Schwartz, H. (1991). *J. Electron Microsc. Tech.*, **17**, 336.
29. Scopsi, I. and Larson, L.-I. (1985). *Histochemistry*, **82**, 321.
30. Danscher, G. (1981). *Histochemistry*, **71**, 81.
31. Bienz, K., Egger, D., and Pasamontes, L. (1986). *J. Histochem. Cytochem.*, **34**, 1337.
32. Danscher, G. (1981). *Histochemistry*, **71**, 1.
33. Scopsi, L. and Larson, L.-I. (1986). *Med. Biol.*, **64**, 139.
34. Allen, R. D., Zacharski, L. R., Widirstky, S. T., Rosenstein, R., Zaitlin, L. M., and Burgess, D. R. (1979). *J. Cell Biol.*, **83**, 126.
35. Kachar, B. (1985). *Science*, **227**, 766.
36. Goodman, S. L., Simmons, S. R., Cooper, S. L. and Albrecht, R. M. (1989). *Proc. Soc. Biomater.*, **15**, 201.
37. Malecki, M., Goodman, S. L., and Albrecht, R. M. (1989). *Proc. EMSA*, **47**, 1002.
38. Miyokawa, T., Norioka, S., and Goto, S. (1988). *Proc. EMSA*, **46**, 978.
39. Simmons, S. R. and Albrecht, R. M. (1989). In *The Science of Biological Specimen Preparation Scanning Microscopy Supplement 3*), (ed. R. M. Albrecht and R. L. Ornberg), pp. 27–34. Scanning Microscopy International, Chicago.
40. Simmons, S. R., Pawley, J. B., and Albrecht, R. M. (1990). *J. Histochem. Cytochem.*, **38**, 1781.
41. Anderson, T. F. (1951). *Trans. New York Acad. Sci.*, **13(II)**, 130.
42. Ris, H. (1985). *J. Cell Biol.*, **100**, 1474.

43. Albrecht, R. M. and MacKenzie, A. P. (1975). In *Principles and Techniques of Scanning Electron Microscopy: Biological Applications*, Vol. 3, (ed. M. A. Hayat), pp. 109–53. Van Nostrand Reinhold, New York.
44. Hayat, M. A. (1989). *Principles and Techniques of Electron Microscopy: Biological Applications*, 2nd edn. Macmillan, London.
45. Franks, J., Clay, C. S., and Peace, G. W. (1980). *Scanning Electron Microsc.*, **1980, I**, 155.
46. Wildhaber, I., Gross, H., and Moor, H. (1985). *Ultramicroscopy*, **16**, 312.
47. Müller, M. and Hermann, R. (1990). *Proc. XIIth ICEM Meeting*, 4.
48. Yonezawa, A., Takeuchi, Y., Kano, T., and Hiroshi, I. (1990). *Proc. XIIth ICEM Meeting*, 396.
49. Hermann, R., Pawley, J., Nagatani, T., and Muller, M. (1988). *Scanning Electron Microsc.*, **11**, 1215–30.
50. Peters, K. R. (1985). *Scanning Electron Microsc.*, **IV**, 1519.
51. Park, K., Albrecht, R. M., Simmons, S. R., and Cooper, S. L. (1986). *J. Colloid Interface Sci.*, **111**, 197.
52. Park, K., Park, H., and Albrecht, R. M. (1989). In *Colloidal Gold: Principles, Methods and Applications*, Vol 1. (ed. M. A. Hayat), pp. 490–514. Academic Press, San Diego.
53. Horisberger, M. and Rosset, J. (1977). *J. Histochem. Cytochem.*, **25**, 295.
54. Autrata, R. (1990). *Proc. XIIth ICEM Meeting*, 376–7.
55. Hermann, R., Schwarz, H., and Mueller, M. (1991). *J. Struct. Biol.*, **107**, 38.
56. Pawley, J. B. (1992). In *Advances in Electronics and Electron Physics*, Vol. 83, (ed. P. Hawkes), pp. 203–73. Academic Press, New York.
57. Albrecht, R. M., Simmons, S. R., and Mosher, D. F. (1990). In *Fibrinogen 4, Current Basic and Clinical Aspects*, (ed. M. Matsuda, S. Iwanaga, and A. Henschen), pp. 87–92. Elsevier, New York.

9

In situ hybridization

J. T. DAVIES

1. Introduction

In situ hybridization—the localization of specific mRNA or DNA species in tissues, cells or chromosomes using nucleic acid probes—has become an increasingly valuable research technique since it was first proposed and demonstrated in the late 1960s (1).

1.1 What is *in situ* hybridization?

Nucleic acid hybridization techniques rely upon the fact that complementary sequences of single-stranded nucleic acid species spontaneously re-anneal, under appropriate conditions, to form a double-stranded hybrid (duplex). Since it is possible, indeed almost trivial nowadays, to incorporate isotopic or non-isotopic markers into nuclei acids, either enzymatically or photo-chemically, we can use these labelled nuclei acid molecules as probes to study the distribution or level of expression of specific genes or gene transcripts.

   In much the same way that a protein must be isolated and purified before a specific antibody against it may be produced, the synthesis of a specific nucleic acid probe requires the isolation of an individual gene or gene transcript. cDNA libraries are the commonest source of such material. A cDNA library consists of representatives of all the mRNA species present in a particular tissue at a given time—a copy of each species being present in an individual replicative form. From such a library, genes specific to a given developmental stage or showing homology to genes of known function or distribution may be isolated and further characterized. Such cDNA clones provide the basis for *in situ* hybridization.

   Most studies of gene expression involve the hybridization of a specific nucleic acid probe to either genomic DNA (Southern blotting (2)) or mRNA (Northern blotting) which has been isolated from homogenized tissue, separated by electrophoretic methods, and transferred to a stable matrix such as nitrocellulose or, more recently, nylon (3, 4). Whilst these techniques can provide a great deal of information about the temporal aspects of gene expression they are limited by the crude spatial discrimination of the technique. It is possible to compare the expression of genes in, for example, the roots of a plant relative to the shoot, whereas *in situ* hybridization allows the precise cellular or subcellular localization of these genes/gene transcripts.

1.2 Why do it?

Whilst conventional immunocytochemical techniques provide information on the spatial patterns of protein distribution they do not allow the underlying changes in mRNA levels to be studied. A protein may persist in the cell long after its translation and long after the concentration of the mRNA encoding it has declined. *In situ* hybridization—where the mRNA species encoding a particular protein may be localized with high resolution both spatially and temporally—provides us with more information. In conjunction with immuno-cytochemical analyses it enables the precise nature of gene expression to be determined, providing information on both the spatial and temporal patterns of expression and on the mechanisms of gene regulation. *In situ* hybridization may also be used to map genes to specific chromosomes (5, 6), to study the integration of viral sequences (7), and to elucidate the organisation of nuclear structures (8). Current techniques enable the localization of specific nucleic acid species in a wide range of tissues and at almost any scale; from ultrastructural to macro/tissue. The recent availability of compatible non-isotopic multiple labelling techniques enables the localization of several genes or gene transcripts in the same tissue or chromosomal spread, greatly enhancing the utility of the system.

Sensitivity, always considered a problem with tissue localization in the past, is now extremely good in optimized systems and the localization of single-copy genes and of mRNA transcripts present at less than ten copies per cell is now possible (9). Indeed, since mRNA species present at low levels are often expressed in only a few cells, the sensitivity of *in situ* hybridization may exceed that of Northern hybridization where the target mRNA is necessarily diluted by the relatively large scale homogenization of tissue. The technique of tissue printing—where the tissue is sectioned without fixation and the cut surface placed directly on to a nitrocellulose membrane (10)—combines the spatial resolution of *in situ* hybridization with the convenience of filter hybridization.

1.3 Practical considerations

In situ hybridization has an undeserved reputation as a fickle and technically demanding technique. In fact the practical considerations are little different from those of conventional immunocytochemistry—the nucleic acid probe being functionally equivalent to the primary antibody (*Figure 1*). It is, however, a more difficult procedure to optimize, owing to the large number of variables involved, and each new tissue type or fixation protocol must be re-evaluated. A brief explanation of technical terms specific to molecular biology is given in the Appendix at the end of this chapter.

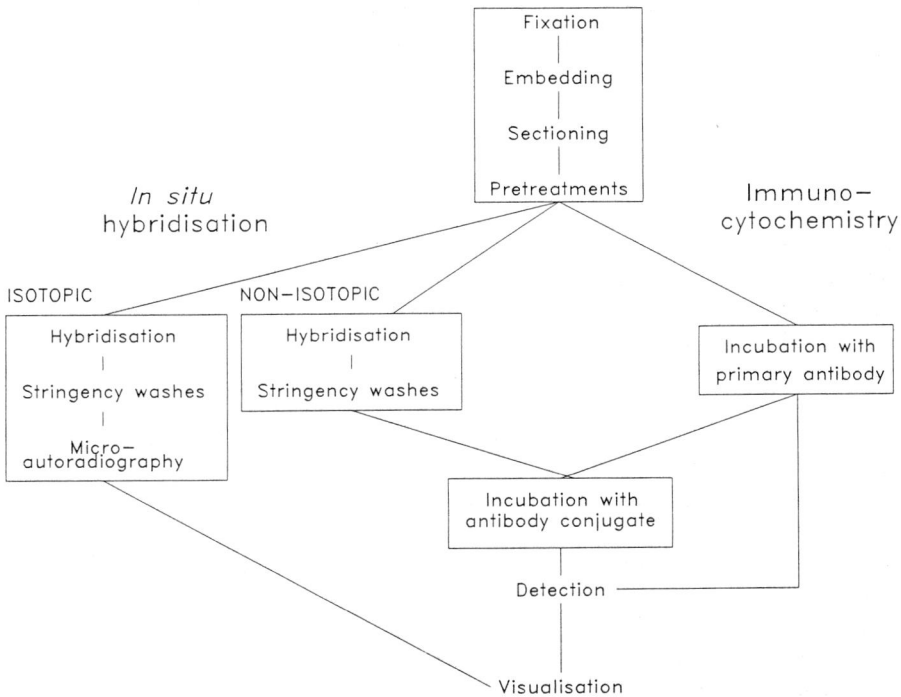

Figure 1. Flow diagram illustrating the similarities between conventional immunocyto-chemistry and *in situ* hybridization.

With *in situ* hybridization, unlike conventional filter hybridization, we have no control over the target concentration. We are able, however, to control a number of variables affecting both the rate and extent of hybrid formation, these include the temperature, probe concentration, probe size, NaCl concentration, formamide concentration, and the presence of dextran sulphate.

1.3.1 Hybridization theory

The extent and rate of the re-annealing process (hybridization)—in which single-stranded species combine to form double-stranded species—is primarily determined by the relative concentration of the reacting species, the temperature, and the ionic strength. Under conditions of low stringency (low temperatures, ionic strength > 0.3 M) double-stranded hybrids may be formed between species of low complementarity—with sequences that differ extensively (have little homology). Under highly stringent conditions, however, only those hybrids formed between species with almost exact complementarity (100% homology) will remain as hybrids.

The maximal rate of nucleic acid hybridization in solution (at ionic strengths above 0.4 M NaCl) occurs at 25 °C below the T_m of the duplex, where T_m is the temperature at which half of the hybrids are dissociated. For probes longer than 50 nucleotides an empirical fomula has been derived to calculate the T_m (11):

$$T_m \text{ (°C)} = 81.5 \text{ °C} + 16.6 \log M + 0.41 \text{ (\%G + \%C)} - 500/n \; 0.61 \text{ (\% formamide)}$$

where M is the ionic strength of the hybridization solution (molar) and n is the length of the shortest chain in the duplex (bp).

Under normal hybridization conditions (0.9 M NaCl) and using a probe of 100 bp with an averge G + C content (40%) the $T_m = 93.7$ °C in the absence of formamide and 63.2 °C in the presence of 50% formamide.

With oligonucleotide probes the base composition has a pronounced effect upon the T_m of the duplex. Under normal hybridization conditions the T_m of the duplex may be estimated using the following formula:

$$T_m \text{ (°C)} = 4 \text{ (number of G/C bases)} + 2 \text{ (number of A/T bases)}.$$

The computer program OLIGO (12) may be used to determine more accurately optimum hybridization conditions for oligonucleotide probes. The use of oligonucleotide probes is difficult since the melting point of the duplex is low and the potential for binding to non-target sequences very high—they can, however, be used to discriminate between closely related targets.

The melting temperature of the probe:target duplex decreases by about 1 °C for every 1% of mismatched base pairs. The incorporation of haptens, such as biotin, into nucleic acids also affects hybrid stability and reduces the T_m by about 5 °C. It might be expected that hybridization should take place at temperatures as close to the expected T_m of the hybrid as possible (highly stringent). A more common approach, however, is to perform the hybridization under conditions of low stringency—promoting hybrid formation—and then to wash off non-specifically bound probe in a series of post-hybridization washes of progressively higher stringency. Washing at low salt and higher temperatures dissociates hybrids formed between the labelled probe and a target which is not homologous.

The conditions under which hybrids are formed in tissue sections, where only a limited section of the probe is able to hybridize to the fixed target (see *Figure 2*), mean that stringency washes used in Southern or Northern blotting experiments are inappropriate to *in situ* hybridization. When utilizing non-isotopic detection techniques, where sensitivity is inherently lower than isotopic detection and where single probe molecules bound non-specifically are not detected, because of the lower sensitivity, the post-hybridization washes may be performed at very low stringency.

1.3.2 RNase-free environment

RNA molecules, whether as a probe or target, are extemely labile and very susceptible to degradation by RNases which target mRNA much less so than

Labelled probe

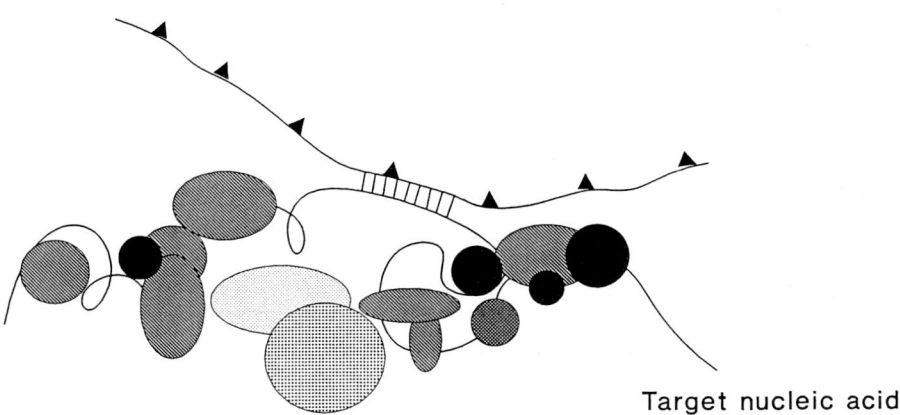

Target nucleic acid

Figure 2. Simplified representation of mRNA species within the cross-linked matrix of a tissue section, illustrating the limited region to which hybridization can occur.

RNA probes because of the fixation of the former. RNase A, widely used in molecular biology, is found in skin secretions, and is virtually indestructible (it is not destroyed by autoclaving alone and may re-nature following phenol extraction).

Equipment to be used for *in situ* hybridization experiments involving either detection of an mRNA target, or utilizing an RNA probe, should be sterile and treated to ensure the complete absence of RNases. RNase-free microcentrifuge tubes and micropipette tips are commercially available (Bioquote Ltd UK) and should be used where possible—even tips not certified RNase-free are generally safe provided they have never been handled without gloves. Solutions may be rendered RNase-free by treatment with the strong acylating compound diethylpyrocarbonate (DEPC) (see *Protocol 1*). Solutions containing Tris salts, other primary amines, and thiol group-containing compounds react with DEPC and should not be treated; Tris solutions should be made up in water previously treated with DEPC.

Glassware may conveniently be rendered RNase-free by baking at 200 °C for 4 h. Any apparatus which cannot be baked, or autoclaved in the presence of DEPC, should be rinsed with a freshly prepared solution of 3% hydrogen peroxide followed by a methanol rinse.

Protocol 1. Treatment of solutions with DEPC

1. Prepare solution to be treated and stir using a magnetic stirrer.

2. Add DEPC[a] to a concentration of 0.1%[b].

3. Shake vigorously then release pressure.

Protocol 1. *Continued*

4. Stir at room temperature overnight.[c]

5. Autoclave (121 °C, 20 min).

---

[a] DEPC is a suspected carcinogen and should be handled in a fume-hood. The bottle may develop internal pressure so exercise due care when opening.
[b] Some workers use at 0.2%.
[c] Alternatively a 3 h incubation with shaking at 37 °C suffices.

---

2. Sample preparation

2.1 Fixation and embedding

Tissue fixation is crucial to the success of the *in situ* localization. It must:

- prevent loss of the target nucleic acids
- preserve tissue morphology
- allow probe penetration

Many fixation protocols, both crosslinking aldehyde types and predominantly precipitative fixatives such as FAA, have proved suitable for *in situ* hybridization. The choice of fixative is tissue dependent and optimum fixation conditions must be determined empirically for each tissue to be studied.

Good results have been obtained with plant material fixed overnight in PBS-buffered 4% paraformaldehyde and subsequently embedded in wax (ParaPlast Plus). Embedding in polyethylene glycol (PEG 1000 or PEG 1500) gives vastly improved morphology but is unsuitable for certain tissues which dissociate following sectioning. In general, a fixation and embedding protocol which has proved suitable for conventional immunocytochemical studies will also serve well, or at least provide a good starting point, for *in situ* hybridization. Fixation and dehydration at temperatures below ambient (0–4 °C) may help reduce degradation of nucleic acids and is recommended.

2.2 Attachment of tissue sections

Since *in situ* hybridization requires the prolonged incubation of tissue sections at relatively high temperatures and in the presence of formamide, it is essential that the sections be firmly adhered to the glass slides. A variety of slide-coating procedures have been successfully utilized for *in situ* hybridization, including gelatin/chrome alum subbing, poly-L-lysine coating, and silanizing treatments. Gelatin-subbed slides occasionally bind the enzyme conjugates involved in certain non-isotopic detection procedures leading to high backgrounds and poly L-lysine-coated slides have an affinity for gold colloids restricting their use when IGSS detection techniques are employed. Slides silanized with TESPA (13) (see *Protocol 2*) exhibit good section adherence alongside very low non-specific binding and their use is recommended for all forms of *in situ* hybridization.

Protocol 2. Preparing TESPA-coated slides

1. Dip clean slides[a] for 10 sec in a 2% solution of 3-aminopropyl triethoxysilane (TESPA) in acetone.

2. Rinse slides briefly in acetone twice.

3. Rinse with DEPC-treated distilled water.

4. Allow to air-dry and store desiccated until required.[b] (Slides are stable for several months.)

---

[a] It is not normally necessary to wash the slides (we use BDH twin-frost premium microscope slides) prior to treatment—indeed washing the slides may result in uneven coating.

[b] The treatment with TESPA probably inactivates any RNase activity associated with the slides but, if desired, the treated slides may be baked (200 °C, 4 h) to ensure the absence of nuclease activity.

2.3 Checking retention of target nucleic acids

Staining of the total nucleic acids within the tissue sections enables rapid confirmation of their retention following fixation and subsequent pre-treatments. It also allows the easy comparison of different protocols. Acridine orange (a nucleic acid intercalating dye) is commonly used for this purpose (see *Protocol 3*) (14). If problems are encountered with tissue autofluorescence masking the acridine orange signal, a common problem with plant tissues and aldehyde fixation, staining with methyl green/pyronin Y is a very acceptable alternative.

Protocol 3. Acridine orange staining of nucleic acids in tissue sections

1. Incubate sections[a] in a 0.5 mg/ml solution of acridine orange dissolved in 0.2 M glycine–HCl pH 2.0 for 30 min[b] in darkness.

2. Wash twice, 15 min each wash, with 0.2 M glycine–HCl pH 2.0.

3. Mount in non-fluorescent mountant (e.g. Citifluor).

4. View under blue (470 nm) excitation fluorescence. DNA fluoresces yellow/green, RNA fluoresces orange/red.

---

[a] A useful control consists of tissue pre-treated with DNase-free RNase A (see *Protocol 9*).

[b] For sections embedded in resin the staining time may need to be extended to 1 h to ensure penetration.

2.4 Pre-treatments

The conditions under which samples are incubated with all the solutions used in the hybridization and detection procedures are determined by the volume required and the cost of the reagent. 200 ml polyethylene slide boxes (TAAB, UK) are convenient for washes; all other incubations are performed in a humid chamber with the solution covering the samples. A coverslip is used only when the incubation is prolonged or the reagent expensive.

Fixed, sectioned material (normally at 7–10 μm) requires pre-treatment to permeabilize the tissue and increase probe accessibility. *Protocol 4* is a generalized protocol for wax-embedded sections prior to localization of an mRNA target. Since pre-treatment exposes the labile RNA species to potential degradation it should be performed just prior to the hybridization.

Protocol 4. Sample preparation

Materials

2 × SSC (0.3 M NaCl, 30 mM Na$_3$citrate pH 7.5)
4% paraformaldehyde in PBS
proteinase K (20 mg/ml in water, stored at −20 °C)
proteinase K buffer (10 mM Tris–HCl pH 7.8, 5.0 mM Na$_2$EDTA, 0.5% SDS)

Method

1. Immerse paraffin-embedded sections in Histoclear[a] for 10 min to remove paraffin.
2. Rinse twice for 5 min in 100% ethanol.
3. Air-dry at room temperature.
4. Rehydrate in 2 × SSC.[b]
5. Immerse slides for 15 min in 0.3% Triton X-100 in 2 × SSC at room temperature.[c]
6. Wash slides twice for 5 min in 2 × SSC.
7. Incubate with proteinase K[d] at 100 μg/ml in proteinase K buffer for 10 min at room temperature.
8. Rinse with 2 × SSC.
9. Refix in 4% paraformaldehyde in PBS for 5 min at room temperature.[e]
10. Rinse with 2 × SSC.
11. Acetylate[f] (see *Protocol 5*).
12. Rinse with distilled water.
13. Air-dry.

a Histoclear is a mixture of purified oils specially prepared to remove wax from preparations. Xylene, toluene, or 1,1,1-trichloroethane (Inhibisol) may be used equally successfully but may carry much greater health risks than those associated with Histoclear.

b TBS or PBS may be used in place of 2 × SSC.

c This step is an absolute requirement for animal tissues only and may be regarded as optional for plant sections and chromosomal spreads.

d This step is crucial to the success of the hybridization with some tissues. Pronase (predigested for 1 h at 37 °C to remove nucleases) may be substituted for proteinase K. The optimum protease concentration, reaction buffer, and incubation time must be determined empirically for each tissue and fixation method. Proteinase K is generally used at 10–100 μg/ml at 37 °C for ± 10 min.

e Fixes tissue and prevents possible target diffusion.

f Acetylation of the tissue can reduce non-specific electrostatic binding of the probe to the tissue sections—for most tissues its use can be regarded as optional.

Protocol 5. Acetylation of samples

Materials

acetic anhydride
0.1 M triethanolamine pH 8.0

Method

1. Place slides in a slide rack and immerse in a large excess of 0.1 M triethanolamine pH 8.0.

2. Place on a magnetic stirrer and stir sufficiently vigorously to circulate the solution without damaging the slides.

3. While stirring, add acetic anhydride to a concentration of 0.25% and incubate for 10 min at room temperature.

4. Wash with distilled water for 1 min.

5. Allow slides to air-dry.

---

3. Types of probe—a comparison

Nucleic acids may exist as double-stranded (ds) molecules, held together by complementary base-pairing, or as single-stranded (ss) molecules. The mRNA species present *in vivo*, which are translated into protein product, exist as single-stranded 'sense' strands. It is possible *in vitro* to synthesize both sense transcripts, which will not hybridize to the mRNA sense-strand present within cells, and the complementary anti-sense transcript which will hybridize. The various forms of nucleic acids (dsDNA, ssDNA, ssRNA, or synthetic oligodeoxyribonucleotides (oligonucleotides)) all have utility as probes in *in situ* hybridization experiments. Each type of probe has its merits (see *Table 1*) and the choice of probe will depend on the nature of the starting material and on the molecular biological experience of the investigator. For more detailed

Table 1. Choice of probe type for *in situ* hybridization

Probe type	Advantages	Disadvantages
RNA	RNA:RNA hybrids have high stability No probe denaturation required No re-annealing during hybridization Probe is strand specific Probe is free of vector	Post-hybridization RNase treatment removes non-hybridized probe Subcloning of probe into dual-promoter vector required Care required to avoid RNase degradation of probe
dsDNA	No subcloning required Good choice of labelling methods Possibility of signal amplification by networking of vector sequences	Probe denaturation required Re-annealing during hybridization Hybrids less stable than with RNA probes
Oligonucleotide	No cloning required No self-hybridization Good tissue penetration Can be constructed from protein sequence data	Limited labelling methods Small size limits amount of label carried Subject to design errors Only short sequence available for hybridization

information the reader is referred to any of the comprehensive molecular biology handbooks (15, 16).

The recent proliferation of easy-to-use kits for the isolation, characterization, and labelling of nucleic acids largely eliminates the need for prior experience of molecular biology. In addition, companies exist (e.g. British Biotechnology Ltd UK) that will synthesize labelled probes to customer specification.

It is now widely accepted that ssRNA probes are the best choice for most *in situ* applications—they allow the use of the sense strand as a control, are of defined length, and, being single-stranded, do not self-anneal during hybridization. In addition, the signal-to-noise ratio obtainable using ssRNA probes is significantly higher than that obtained with dsDNA probes since the RNA:DNA or RNA:RNA duplexes formed are more stable—allowing the use of more stringent hybridization conditions—and RNase A treatment following the post-hybridization washes may be used to degrade any remaining RNA probe bound non-specifically. This combination of factors is reported to result in an eight fold increase in sensitivity relative to dsDNA probes (17).

4. Labelling of nucleic acid probes

4.1 Isotopic and non-isotopic—a comparison

4.1.1 Isotopic probes

Currently, the vast majority of filter hybridizations are conducted using isotopically labelled (normally ^{32}P) nucleic acid probes. Such probes provide both high sensitivity and the ability to wash the filter at progressively higher stringencies before re-exposing to autoradiographic emulsion. The advantages of isotopically labelled probes for filter hybridization do not apply to *in situ* hybridization, where resolution is often more important than sensitivity and where it is not possible to remove bound probe at higher stringency once the sections have been dried and dipped in microautoradiographic emulsion.

The radioisotopes suitable for use with *in situ* hybridization are listed in *Table 2* along with their respective advantages and normal application.

Table 2. Choice of radiolabel for isotopic *in situ* hybridization

Label	Application	Advantage	Half-life	Resolution
^{32}P	Macro-scale	Very rapid analysis	14.3 days	20–30 μm
^{35}S	Cellular	Short exposure	87.4 days	10–15 μm
^{125}I	Sub-cellular	High sensitivity	60.0 days	1–10 μm
^{3}H	Sub-cellular	High resolution	12.4 years	0.5–1.0 μm

4.1.2 Non-isotopic probes

The use of non-isotopic markers for labelling nucleic acids has increased substantially over the past decade (18–21). They provide a number of advantages over isotopic markers:

(a) advantages:

- labelled probe may be stored (> 6 months at −20 °C)
- high resolution
- rapid results
- multilabelling possible
- safety

(b) disadvantages:

- lower sensitivity
- greater potential for non-specific interactions

Until recently, biotin-labelled probes provided the experimenter with the highest sensitivity and with the widest choice of localization techniques. Endogenous biotin within tissue samples has, however, limited the utility of the system when applied *in situ*. The problems of endogenous biotin may be avoided by using an alternative label such as digoxigenin. Kits for the incorporation of the digoxigenin moiety into nuclei acids and for its subsequent detection are available (Boehringer Mannheim UK, USA) and now provide all the flexibility of the biotin systems (22, 23).

The most commonly used non-isotopic markers and the options for their detection are listed in *Table 3*.

Table 3. Non-isotopic detection systems for *in situ* hybridization

Marker	Detected with	Comments
Biotin[a]	Avidin	Avidin is a glycoprotein High non-specific binding
	Streptavidin	Better than avidin but large molecule Poor tissue penetration
	Anti-biotin	Lower affinity than avidin but better tissue penetration
Digoxigenin[b]	Anti-digoxigenin	No problems with endogenous marker
Direct enzyme (37)		Time-consuming preparation Large conjugates; poor penetration
Direct fluorescence[c]		Very rapid localization Lower sensitivity than indirect methods

[a] The most common problem with biotin, especially with tissue derived from cell suspension cultures, is the presence of endogenous biotin.

[b] Digoxigenin is a steroid derived from foxglove which may be incorporated into nucleic acids and detected using specific antibodies—no problems with endogenous digoxigenin in tissues.

[c] Enzymatically incorporable fluoresceinylated nucleotide analogs (UTP, dUTP, and ddUTP) are available from Boehringer. Resorufin-10-dUTP and hydroxycoumarin-6-dUTP fluorescent nucleotide analogues will be made available from the same source in the near future.

4.2 Controlling probe length

In order to hybridize to nucleic acids within the fixed material the probe must first penetrate the tissue—shorter probes exhibit better penetration but carry less label and are more susceptible to anomalous hybridization. For most work, probes of 200–500 nucleotides prove optimal and these may be derived from all common labelling methods.

Oligonucleotide probes, being very short (generally 20–40 nucleotides, are able to enter tissues easily; their small size, however, also limits the amount of label they can carry.

ssRNA probes derived from dual-promoter plasmids have, by nature of their synthesis, a defined length (24). Such probes may be shortened by controlled alkaline hydrolysis to yield a range of probe lengths more appropriate to *in situ* hybridization (25). The average length of probes synthesized by nick translation (where DNase activity nicks the DNA, creating sites from which the polymerase can incorporate fresh nucleotides) is determined by the ratio of DNase I to DNA Polymerase I, parameters which can be controlled by the experimenter to yield probes of the desired size. The random primed method of DNA labelling (26, 27) allows probe size to be controlled by appropriate choice of hexanucleotide primer concentration. The higher the concentration of the primers, the shorter the average distance between priming sites on the template DNA and the shorter the average probe length (28).

Photolabelling (see *Protocol 6*), whilst providing only about 1/10 the sensitivity of enzymatic incorporation methods, is convenient and is applicable to both RNA and DNA preparations. Because of its near universal applicability, and its relative simplicity, this technique can be recommended for all applications not requiring the ultimate in sensitivity or the use of oligonucleotide probes. Photoreactive forms of both biotin (Sigma UK, USA and Boehringer Mannheim UK, USA) and digoxigenin (Boehringer Mannheim UK, USA) are available.

Protocol 6. Photolabelling of nucleic acids using photoreactive biotin

Materials

100% and 70% (v/v) ethanol at −20 °C
4 M LiCl
5 M NaCl
100 mM Tris–HCl, 1.0 mM Na_2EDTA pH 9.0
TE (10 mM Tris–HCl, 1.0 mM Na_2EDTA pH 8.0)
water-saturated 2-butanol
Philips HPLR 400 W lamp (or equivalent)

Method

1. In darkness[a] mix 1 µg DNA or RNA with 3 µg photobiotin (= 1 µl) in a total volume of 10 µl.

2. Place the open tube on ice 10 cm from a Philips HPLR 400 W lamp.

3. Irradiate the nucleic acid–photoreactive biotin mixture for 10 min (RNA labelling) or 15 min (DNA labelling).

4. Add 90 µl 100 mM Tris–HCl, 1.0 mM Na_2EDTA pH 9.0.

5. Add 5 µl 5 M NaCl and mix thoroughly.

6. Extract twice with 100 µl water-saturated 2-butanol.[b]

7. Precipitate the biotin-labelled nucleic acid by adding 10 µl 4 M LiCl and 200 µl pre-chilled (−20 °C) ethanol.

8. Incubate for at least 2 h at −20 °C or 40 min at −70 °C.

9. Centrifuge (12 000g) for 30 min at room temperature.

10. Wash the pellet with 100 µl pre-chilled 70% ethanol.

11. Carefully pipette off excess ethanol and air-dry pellet at room temperature.

12. Redissolve pellet in 50 µl TE.

13. Store the labelled probe at −20 °C.

[a] Or under suitable safelighting (e.g. Ilford 902).
[b] The labelled probe remains in the lower aqueous layer while the unincorporated photoreactive biotin partitions into the upper butanol layer (which turns pink).

---

5. Hybridization *in situ*

5.1 Introduction

Hybridization of the probe with the specimen is considered in this section, whereas localization of the bound probe is considered in Section 6.

5.2 Solutions

The solution in which the nucleic acid hybridization takes place appears, at first sight, to be a complex mixture. Each component has, however, a simple function and the hybridization buffer is similar, in both function and form, to the buffered solutions (containing blocking compounds) employed for antibody incubation in immunocytochemical studies. A typical hybridization buffer is detailed in *Table 4*.

Table 4. Typical *in situ* hybridization buffer

Component	Stock	Final	Function
SSC	20 ×	5 ×	Buffer/[NaCl][a]
Deionized formamide	100%	50%	Lowers T_m of hybrids[b]
Denhardt's solution[c]	50 ×	5 ×	Blocking
Denatured, sheared nhDNA[d]	1 mg/ml	100 µg/ml	Blocking
SDS	20%	2%	Blocking
Dextran sulphate	50%	10%	Volume exclusion[e]
Labelled probe		To ± 1 µg/ml	
DEPC-treated water		To total	

[a] SSPE may be substituted for SSC and is the preferred buffer for hybridizations involving formamide.

[b] By lowering the T_m (melting point) of the hybrid nucleic acids the use of 50% formamide enables hybridization to take place at a lower temperature and consequently with less morphological damage to the tissue.

[c] A blocking solution composed of BSA, PVP, and Ficoll in equal proportions by weight.

[d] Non-homologous (usually herring or salmon sperm) DNA.

[e] By sequestering water and thereby effectively increasing the probe concentration, dextran sulphate increases the rate of hybridization without the concomitant increase in background normally associated with its use in filter hybridizations.

SSC and SSPE are the buffers most commonly used in molecular biology and are equivalent to PBS or TBS in function. Since most nucleases (RNases being notable exceptions) require Mg^{2+} ions as a co-factor, both SSC and SSPE incorporate some degree of metal chelating ability (citrate ions in the case of SSC and Na_2EDTA in the case of SSPE). Since SSPE has the greater buffering capacity it is the preferred buffer in hybridization solutions containing formamide. All the reagents used for *in situ* hybridization should be of Analar grade, dissolved in double-distilled water and sterilized by autoclaving or sterile filtering as appropriate.

Whilst Denhardt's solution (29) and sheared non-homologous DNA may be prepared relatively easily in the laboratory, the small quantities required for *in situ* hybridization, and the necessity that they be RNase-free, make the purchase of commercial preparations (Sigma UK, USA) very attractive.

The sections may be hybridized in a conventional bench-top oven with temperature regulated to ± 0.5 °C; it is not necessary to employ the hot oil immersion methods employed by earlier workers. Hybridization is normally allowed to proceed overnight but the high probe concentrations employed in most non-isotopic hybridization methods mean that hybrid formation is essentially complete after 3–5 h. It is possible, therefore, to perform an entire non-isotopic *in situ* localization in a single, albeit long, working day.

5.3 Hybridization to mRNA target using digoxigenin-labelled dsDNA probe

Protocol 7 details the hybridization of a digoxigenin-labelled dsDNA probe to an mRNA target. The protocol is generally applicable—dsDNA probes labelled isotopically, with biotin or fluorescent nucleotide derivatives—could all be substituted for the digoxigenin-labelled probe used to illustrate the method.

Protocol 7. Hybridization to mRNA target using digoxigenin-labelled dsDNA probe

Materials

deionized formamide[a]
20 × SSC (3 M NaCl, 0.3 M Na citrate pH 7.5)
50 × Denhardt's solution (1% BSA, 1% PVP (polyvinylpyrrolidone) and 1% Ficoll in water)
sonicated, denatured herring sperm DNA (10 mg/ml)
yeast tRNA (10 mg/ml)
dextran sulphate (50% stock, dissolved in DEPC-treated water at ± 60 °C, stored at 4 °C)
20% SDS (sodium dodecylsulphate)

Method

1. Prepare the hybridization solution as follows, adding the components in the order given, to prepare 1.0 ml (sufficient for 20 slides if not pre-hybridized)

   - deionized formamide   500 μl
   - SSC (20 ×)   200 μl
   - Denhardt's solution (50 ×)   20 μl
   - herring sperm DNA (10 mg/ml)   50 μl
   - yeast tRNA (10 mg/ml)   25 μl
   - dextran sulphate (50%)   200 μl

Protocol 7. *Continued*

2. Heat-denature the labelled DNA by heating to 95 °C for 10 min and cool in an ice/NaCl bath for 5 min.

3. Add the labelled, denatured DNA probe (in ± 5 μl) to the hybridization solution to a concentration of 1 mg/ml and mix thoroughly.

4. Pipette 50 μl of the probe solution onto the pre-treated sections taking care to avoid air bubbles.

5. Cover with a siliconized coverslip.[b]

6. Place slides in a humid chamber containing 4 × SSC, 50% formamide.[c]

7. Incubate overnight at 42 °C.

8. Immerse slides in 4 × SSC, 50% formamide at 42 °C until coverslips become detached.

9. Wash slides in a large excess (> 10 ml/slide):

   • 4 × SSC, 0.1% SDS twice for 30 min at room temperature

   • 2 × SSC, 0.1% SDS twice for 30 min at 42 °C

10. Visualize bound probe (see *Protocol 12*).

---

[a] Deionize formamide by stirring with a mixed ion-exchange resin (BioRad AG 501-x8) for 30 min and filter to remove resin and store in small aliquots at −20 °C.

[b] The coverslips are easily siliconized using commercial preparations such as Repelcote (BDH, UK). Some workers seal the coverslips with rubber adhesive but this is inconvenient and unnecessary in my experience. Small squares of Parafilm or of Gelbond (hydrophobic side down) are very effective alternatives.

[c] It is essential that the humid chamber contains a buffer of similar composition to the hybridization solution to avoid volume changes under the coverslip.

5.4 Isotopic (^{35}S) mRNA localization

When isotopically labelled probes are intended for use *in situ*, where high probe concentrations are necessary, the quantity of probe synthesized may be as important as its specific activity. For *in situ* applications, therefore, the probe labelling protocol may be modified by the dilution of the radiolabelled nucleotide with 'cold' nucleotide to provide a probe with a practical balance between specific activity and quantity.

Protocol 8. mRNA localization using ^{35}S-labelled ssRNA probes

Materials

As *Protocol 7*, with the addition of:
20 × SSPE (3.6 M NaCl, 0.2 M sodium phosphate pH 7.7, 0.02 M Na$_2$EDTA)
dithiothreitol (DTT)
7.5 M ammonium acetate
DNase-free RNase A (see *Protocol 9*)

Method

1. Prepare hybridization buffer as follows, adding the components in the order given (10 ml volume; final concentration of reagents in parenthesis):

   - deionized formamide   5 ml   (50%)
   - 20 × SSPE   2.5 ml   (5 ×)

   warm to 37 °C and add

   - dextran sulphate   1 g   (10%)

   mix well until dextran sulphate dissolves and add

   - 50 × Denhardt's solution   1000 μl   (5 ×)
   - 20% SDS   250 μl   (0.5%)
   - 10 mg/ml denatured, sheared herring sperm DNA   100 μl   (100 μg/ml)
   - DTT   154 mg   (100 mM)
   - DEPC-treated distilled water   to 10 ml

2. Keep hybridization buffer at 37 °C until required.

3. Add probe to a final concentration of 50–200 ng/ml.[a]

4. Place 50 μl of the probe solution over the sections.

5. Cover with siliconized coverslips.

6. Incubate overnight at 50 °C in a humid chamber containing 5 × SSPE, 50% formamide.

7. Remove coverslips by immersing slides in 4 × SSC, 50% formamide, 10 mM DTT, 0.1% SDS.

8. Wash twice, 30 min each wash, with excess (10 ml/slide) 4 × SSC, 50% formamide, 10 mM DTT, 0.1% SDS.

9. Wash twice, 30 min each wash, with excess (10 ml/slide) 2 × SSC, 50% formamide, 10 mM DTT, 0.1% SDS.

10. Rinse briefly in 2 × SSC.[b]

11. Incubate slides with 10 μl/ml DNase-free RNase A in 2 × SSC at 37 °C for 15 min.

12. Dehydrate through an ethanol series containing 0.3 M ammonium acetate as follows:

    - 70% ethanol   30 sec
    - 90% ethanol   30 sec
    - 100% ethanol   60 sec
    - 100% ethanol   60 sec

Protocol 8. *Continued*

13. Air-dry at room temperature.

14. Visualize bound probe (see Section 6.2)

[a] Maximum sensitivity is achieved at probe levels just high enough to saturate the target sequences (32). This concentration is proportional to the average probe length—a good starting point is 0.3 μg/ml/kb probe length.
[b] This step removes traces of SDS which may inhibit RNase activity.

5.5 Non-isotopic (fluorochrome) chromosomal localization

The localization of genes in isolated chromosome preparations has proved a useful technique for the low resolution mapping of genes and for studies of chromosomal dislocations. The method may be applied to routinely G-banded chromosomes (34, 35). By using both biotin- and digoxigenin-labelled probes it is possible to localize two or more genes in the same preparation using detection reagents coupled to different fluorochromes. For the detection of two different species, the probe against one would be labelled with biotin, the other with digoxigenin. Following hybridization with both probes (they can generally be hybridized simultaneously) they are easily discriminated by detection with the highly specific fluorescent antibody conjugates available.

The recent availability of several fluorescent nucleotide analogues (Boehringer Mannheim UK, USA), which may be incorporated into nucleic acid probes using conventional techniques, makes the simultaneous detection of multiple targets much simpler—probes against each target species being labelled with a different fluorochrome. Such directly incorporated fluorescent probes have the disadvantage of slightly reduced sensitivity when compared to indirect methods but, if desired, the signal may be amplified using antibodies directed against the fluorescent moiety.

Protocol 9. Preparation of DNase-free RNase A

1. Dissolve pancreatic RNase A at a concentration of 10 mg/ml in 15 mM NaCl, 10 mM Tris–HCl pH 7.5.

2. Heat to 100 °C for 15 min.

3. Allow to cool slowly to room temperature.

4. Aliquot and store at −20 °C.

Protocol 10. Chromosomal localization using a non-isotopic (digoxigenin) system with fluorescent detection

Materials

As *Protocol 7* with the addition of:
DNase-free RNase A (see *Protocol 9*)
0.2 M $MgCl_2$
incubators at 80 °C and 50 °C
fluorescein- or rhodamine-conjugated anti-digoxigenin

Method

1. Prepare meta- or prophase chromosome preparations.

2. Incubate with 100 μl of 100 μg/ml DNase-free RNase A in 2 × SSC under a coverslip for 1 h at 37 °C.

3. Wash three times in 2 × SSC (5 min each wash) and dehydrate through a graded ethanol series.

4. Incubate with proteinase K at 10 μg/ml in TE.

5. Wash with PBS containing 50 mM $MgCl_2$.

6. Post-fix with 1% paraformaldehyde in PBS containing 50 mM $MgCl_2$ 10 min at room temperature.

7. Wash with PBS.

8. Dehydrate through ethanol series.

9. Prepare hybridization solution as follows:[a]

   - deionized formamide   500 μl
   - 20 × SSC   250 μl
   - dextran sulphate (50%)   200 μl
   - sheared, denatured nhDNA (100 μg/ml)   25 μl

10. Add digoxigenin-labelled probe to hybridization buffer to 1 μg/ml.

11. Place 30 μl of the hybridization solution on the slides, cover with a siliconized coverslip and seal with rubber cement (sealing with cement is necessary in this instance because of the high temperature incubation).

12. Denature the probe and the chromosomes together by incubating at 80 °C for 3 min.

13. Hybridize overnight at an appropriate temperature.

14. Remove rubber cement.

15. Wash off the coverslip in 2 × SSC at room temperature for 5 min.

16. Wash with PBS for 5 min; remove excess buffer.[b]

195

Protocol 10. *Continued*

17. Apply 20 μl of a 1:10 dilution (in PBS containing 1% BSA) of fluorescein- or rhodamine-conjugated anti-digoxigenin antibody.

18. Incubate for 30 min at 37 °C.

19. Wash for 5 min with PBS.

20. Mount in suitable mountant.[c]

21. Observe with a fluorescence microscope.

[a] The hybridization and post-hybridization washing conditions required depend on the probe type and on whether the target sequence is unique or repetitive.

[b] Excess buffer may be removed by wiping gently with a soft tissue. The preparations should not be allowed to dry completely since this produces background.

[c] Such as a glycerol–paraphenylenediamine mixture (0.1% (w/v) paraphenylenediamine in 1.0 mM Na phosphate pH 8.0, 15 mM NaCl, 50% glycerol).

5.6 *In situ* hybridization for electron microscopy

The requirements of localization of mRNA transcripts at the electron microscope (EM) level are similar to those of EM immunolocalization of protein antigens. There is a requirement for the preservation of a reasonable amount of structural detail and also the RNA or DNA of interest must be preserved. This requires that the tissue is fixed prior to further processing, which may involve either hybridization followed by sectioning or, more usually, sectioning followed by hybridization. To facilitate probe penetration the tissue may be exposed by removal of the matrix used for support during sectioning or, as with much EM immunocytochemistry, the hybridization may be successfully performed on sections of resin-embedded tissue (35, 36).

The choice of whether to section prior to hybridization or vice versa is determined by the nature of the tissue being studied. For monolayers, epidermal cells, suspension cells, and plant protoplasts it is likely that a mild fixation, hybridization, and staining (and embedding and sectioning if required) will be suitable. For cells within a multicellular mass, however, mild fixation, sectioning, hybridization, and staining may be more appropriate.

Protocol 11. Localization of mRNA species at the electron microscope level

Materials

As *Protocol 7* with the addition of:
SC buffer (50 mM Pipes, pH 7.2 , 0.5 M NaCl, 0.5% Tween 20)
TE (10 mM Tris–HCl, 1.0 mM Na$_2$EDTA pH 8.0)
rabbit anti-biotin
goat anti-rabbit 15 nm colloidal gold

Method

1. Fix tissue in 2% glutaraldehyde in 50 mM Pipes (pH 7.0) on ice for 60 min.
2. Wash in 50 mM Pipes (pH 7.0).
3. Dehydrate through an ethanol series from 20% to 70% ethanol.
4. Infiltrate and embed in LR Gold resin (London Resin Company, UK).
5. Collect ultrathin sections on to Pioloform-coated gold grids.
6. Treat with proteinase K (1 μg/ml in TE) for 15 min at 20 °C.
7. Rinse with water and air-dry.
8. Prepare hybridization buffer as follows:
   - 50% formamide   500 μl
   - 20 × SSPE   250 μl
   - 100 μg/ml sheared, denatured nhDNA 50 μl
   - 50 × Denhardt's solution   100 μl
   - biotinylated probe   to 2 μg/ml
   - DEPC-treated water   to 1000 μl
9. Invert grids onto a 3.5 μl droplet of hybridization buffer containing the labelled probe.
10. Hybridize overnight at 42 °C in a **humid chamber** containing 50% formamide, 4 × SSC.
11. Rinse grids in 4 × SSC.
12. Wash in 1 × SSC at 52 °C for 2 h.
13. Block with 1% BSA in SC buffer.
14. Incubate with rabbit anti-biotin 1:50[a] in SC buffer for 1 h at room temperature.
15. Wash thoroughly with SC buffer.
16. Incubate with goat anti-rabbit 15 nm colloidal gold conjugate 1:10 dilution[a] in SC buffer for 1 h at room temperature.
17. Wash thoroughly with SC buffer.
18. Rinse with distilled water then air-dry.
19. Stain with uranyl acetate and lead citrate.
20. Observe with electron microscope.

[a] The dilutions given are for specific products—optimum dilutions should be determined empirically based on the manufacturer's recommendations. Whilst this protocol employs a biotinylated probe, digoxigenin-labelled probes—in conjunction with the appropriate antibodies—are equally well suited.

6. Visualization of bound probe

6.1 Localization of bound non-isotopically labelled probe

Bound probe labelled with biotin or digoxigenin may be localized and visualized by using antibodies conjugated to an enzyme (alkaline phosphatase or peroxidase), a fluorochrome, or colloidal gold. The relative merits of each detection technique are summarized in *Table 5*. The choice of detection method will depend on the sensitivity and, more importantly, the spatial resolution required. Amersham UK, USA and BioCell UK have recently made available detection reagents (anti-biotin, streptavidin, and, from BioCell UK, anti-digoxigenin) conjugated to 1 nm colloidal gold. These are reported (31, 32) to offer improved sensitivity, better tissue preparation, and higher signal-to-noise ratios than 5 or 10 nm gold conjugates. Kits are available (British Biotechnology Ltd UK) which contain all the reagents required for the enzymatic or fluorescent detection of biotin- or digoxigenin-labelled probes.

Table 5. Choice of non-isotopic detection methods for light microscope methods

	Fluorochrome	Enzyme	Colloidal gold
Sensitivity	++	+++	++[a]
Resolution	++	+	+++
Permanence	+	++	+++
Speed	+++	+	++
Multi-labelling	+++	++	+[b]
Main application	Cellular	Tissue	Tissue/cell

[a] When silver-enhanced and viewed with epi-polarized illumination.
[b] Colloidal gold is, of course, highly suitable for electron microscopic studies.

Protocol 12 details the localization of a digoxigenin-labelled probe using an alkaline phosphatase(AP)-conjugated antibody in combination with a substrate solution giving rise to an insoluble, coloured precipitate. NAMP/Fast Red TR are the preferred substrates for alkaline phosphatase for most applications: the red product is insoluble in water, contrasts effectively with counterstains such as light green, and is easily photographed. In addition, this substrate combination is less prone to non-specific precipitation than the NBT/BCIP combination usually recommended.

Protocol 12. Detection of bound probe

Materials

TBS (100 mM NaCl, 10 mM Tris–HCl pH 8.0)
normal sheep serum
Triton X-100
anti-digoxigenin–AP conjugate (Boehringer Mannheim USA)

naphthol AS-MX phosphate (NAMP) (50 mM in DMF)
Fast Red TR salt (200 mM in 70% DMF)
levamisole
TE (10 mM Tris–HCl, 1.0 mM Na_2EDTA pH 8.0)

Method

1. Rehydrate sections with TBS.[a]

2. Incubate sections in TBS containing 2% normal sheep serum and 0.3% Triton X-100 for 30 min at room temperature.

3. Dilute anti-digoxigenin–AP conjugate 1:500 with TBs containing 2% normal sheep serum and 0.3% Triton X-100 and use immediately.

4. Cover the sections with the diluted anti-digoxigenin antibody solution ($\pm$ 100 µl/slide) and incubate in a humid chamber for 2 h at room temperature.

5. Wash slides for 1 h in five changes of TBS.

6. Prepare substrate solution by adding 10 µl NAMP stock and 10 µl Fast Red TR stock to 5 ml TBS. Optionally levamisole[b] may be added to a final concentration of 50 mM.[c]

7. Cover sections with substrate solution and incubate in darkness until colour has developed sufficiently. Colour development may occur within 10 min or take up to 16 h. The substrates begin to precipitate non-specifically after 16 h so for extended incubations the sections should be rinsed and covered with fresh substrate. Elevated temperatures increase the rate of colour formation but also increase background markedly.

8. Immerse slides in TE to halt the colour reaction.

9. Rinse sections with distilled water.

10. If desired, counterstain sections with an appropriate stain such as light green.

11. Allow sections to air-dry at room temperature.

12. Mount in suitable mountant.[d]

[a] TBS is used in preference to PBS throughout this protocol since phosphate ions are potent competitive inhibitors of alkaline phosphatase activity.

[b] Levamisole inhibits endogenous alkaline phosphatase activity in most tissues without inhibiting the activity of the calf intestinal phosphatase used as the detection enzyme.

[c] It is worthwhile combining an aliquot of the prepared substrate solution with any remaining anti-digoxigenin–AP conjugate. Colour development occurs within seconds and provides a clear indication that the substrate solution is fully functional.

[d] Loctite GlassBond is an excellent mountant for many tissues, especially plant sections where its refractive index closely matches that of the cell walls and renders them almost invisible under bright field illumination. Place a small drop of GlassBond on the sections and cover with a clean glass coverslip. When the GlassBond has covered all the sections and extends to the edges of the coverslip expose the slides to unfiltered daylight for a couple of minutes to cross-link the anaerobic adhesive.

6.2 Microautoradiography

For high resolution localization of the bound, isotopically-labelled probe the slides must be dipped in an autoradiographic emulsion and allowed to expose for several days (^{35}S) or weeks (^{3}H). Suitable emulsions are available from Amersham (UK, USA) (LM-1), Kodak USA (NTB-2), and Ilford UK (K5). We routinely use Amersham's LM-1 emulsion with good results. The emulsions are supplied with comprehensive instructions for their use and if these are followed carefully no problems should be encountered.

Strong signals are readily visualized with conventional bright field illumination (see *Figure 3*) but dark field or epi-polarizing illumination is recommended for the observation of weaker signals. If desired, sections may be counterstained prior to mounting.

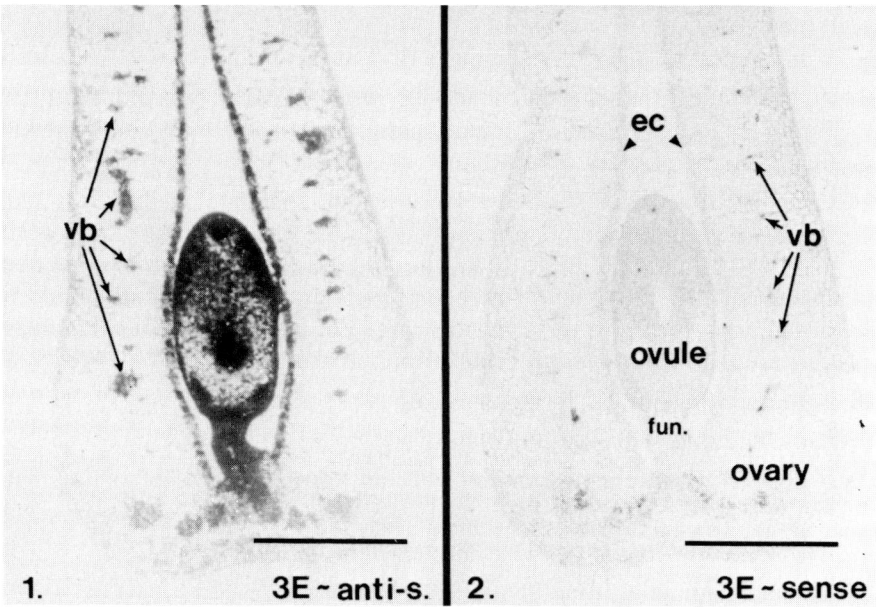

Figure 3. Localization of the mRNA encoding a thiol protease in transverse sections of pea fruit, derived from emasculated flowers, using a ^{35}S-labelled ssRNA probe. The positive result from the anti-sense probe (1), with a strong signal in the ovule, vascular bundles, and endocarp, is shown alongside the sense control (2) (vb, vascular bundle; ec, endocarp) Scale bar 1.7 mm.

7. Controls

To enable meaningful interpretation of the results of *in situ* hybridization experiments requires the use of careful controls—the number of factors affecting the procedure make the analysis of results particularly difficult. The controls generally required are summarized in *Table 6*.

As mentioned earlier, the best negative control probes are the sense-strand RNA molecules derived from dual-promoter vectors. It should be remembered, however, that the bacteriophage promoters T3, T7, and SP6 incorporate nucleotide analogues, such as biotin-11-dUTP, with differing efficiencies. If the sense and anti-sense probes are derived from different promoters, as is common, this must be taken into account when analysing the results.

In situ hybridization experiments are often undertaken with some prior knowledge as to the location or level of expression of a gene or gene transcript. It is easy for such information to prejudice the interpretation of results—it is important to remember that a negative result may be due not to problems with the experiment, but to a genuine absence of the target in those cells at that time verified by the positive control.

Table 6. Controls for *in situ* hybridization

Non-specific probe binding	Sense RNA probe
	Non-homologous probe
	Vector sequence
	Nuclease digestion of target[a]
	No denaturation of DNA target
	Negative tissue
Detection reagents	No probe
	Tissue only
	Blank slide
Reproducibility	Positive tissue
	Positive control probe[b]
Target distribution	Immunocytochemistry[c]

[a] Care must be taken to ensure complete removal of exogenous nucleases otherwise the probe itself may be digested.
[b] Suitable positive control probes include rRNA genes and labelled oligo dT ($\pm$ 20 nucleotides long—this hybridizes to the poly-A tail present on most mRNA molecules).
[c] Obviously only applicable when an appropriate antibody is available.

---

8. Trouble-shooting

The relatively large number of variables affecting *in situ* hybridization make it a system difficult both to optimize and in which to determine the cause of problems. The commonest problems and suggested solutions are detailed in *Table 7*.

Table 7. Trouble-shooting

Problem	Possible cause	Remedy
Sections fall off	Inappropriate slide coating procedure	Change slide coating procedure
	Insufficient drying down of sections on slides or at too low a temperature	Dry sections on slides for at least 4 h at 50 °C
No signal	No target	Check retention of nucleic acids in tissue
	Probe not labelled	Check probe on dot blot
	Poor probe penetration	Permeabilize tissue and/or reduce probe length
	Stringency of hybridization or subsequent washes too high	Lower stringency (increase (NaCl) or lower temperature)
	Probe or target not denatured	dsDNA targets or probes must be denatured
High background (sections)	Insufficiently stringent washes	Increase stringency (lower (NaCl) or increase temperature)
	Probe concentration too high	Lower probe concentration
	Sections allowed to dry out	Do not allow sections to dry out
	High levels of unincorporated nucleotides	Purify probe further
	Probe GC-rich	Use a more selective probe
	Vector sequences present in probe	Purify probe
	Non-specific binding	Increase blocking and/or acetylate tissue
	^{35}S-labelled probes oxidized	Keep ^{35}S-labelled probes reduced with DTT (10 mM)
High background (sections and slide)	Inappropriate slide coating	Change coating procedure
	Substrate incubation too long or elevated temperature	Monitor colour reaction carefully
	Excessive silver enhancement of gold marker	Monitor enhancement carefully
	Poly-L-lysine-coated slides used with colloidal gold	Change coating procedure
	Inadequate de-waxing of sections	Use fresh solvents

9. Conclusions—the future

There are a great many more nucleic acid probes available than there are antibodies against specific proteins. This, coupled with the relative ease with which nucleic acids may be characterized, makes their application to localization *in situ* very attractive. Indeed, molecular biological techniques allow the interconversion of mRNA, cDNA into protein; RNA derived from a cDNA clone may be translated into a protein product against which antibodies can be developed.

Polymerase chain reaction (PCR) has revolutionized many fields of biology, enabling the selective amplification of nucleic acids. Its application to studies *in situ* has been proposed and attempted with limited success. In future the technique may well have an important role to play, increasing the sensitivity of *in situ* hybridization and allowing the localization of gene transcripts present at only single copy per cell. In addition, advances in the quantitation of *in situ* hybridization signals will provide yet more information.

In future I hope that *in situ* hybridization will be seen for what it is—a powerful and synergistic union of microscopy and molecular biology.

Acknowledgements

I would like to thank Nick Harris for his advice and encouragement. Amersham International plc UK, BioCell Research Laboratories UK, and Boehringer Mannheim UK kindly gave permission to reprint certain of their published protocols. Thanks are also due to Lesley Edwards and Jacqueline Spence for their constructive criticism and for permission to use unpublished data.

Appendix. Brief explanation of terms commonly used for *in situ* hybridization

bp	Base pairs; essentially the number of nucleotides in a single strand of a single nucleic acid molecule—the commonest unit of nucleic acid length
cDNA	Copy DNA (complementary DNA)
Denhardt's solution	A blocking solution comprising 1% BSA, 1% Ficoll 400, and 1% polyvinylpyrrollidone
DEPC	Diethylpyrocarbonate—a powerful acylating compound used to render solutions/glassware RNase-free and thus prevent degradation of RNA
ds	Double-stranded (duplex)

kb	Kilobase—a unit of nuclei acid length corresponding to 1000 bp (see above)
mRNA	Messenger RNA
nh	Non-homologous—usually applied to salmon or herring sperm DNA which does not hybridize to most tissues/sequences and may, therefore, be used as a blocking reagent
PBS	Phosphate-buffered saline
RNase A	Ribonuclease A—degrades RNA, whether probe or target. It is present in skin secretions and is virtually indestructible—it is not inactivated by autoclaving alone and can renature following phenol extraction
RNasin	A proteinaceous inhibitor of many ribonucleases
rRNA	ribosomal RNA
ss	Single-stranded
SSC	Standard saline citrate—a ubiquitous buffer in molecular biology
SSPE	As SSC but less commonly employed
stringency	A term used to express the relative discrimination between related nucleic acid species achieved by varying the post-hybridization washing conditions
TBS	Tris-buffered saline

References

1. Gall, J. G. and Pardue, M. L. (1969). *Proc. Natl. Acad. Sci. USA*, **63**, 378.
2. Southern, E. M. (1975). *J. Mol. Biol.*, **98**, 503.
3. Thomas, P. (1980). *Proc. Natl. Acad. Sci. USA*, **77**, 5201.
4. Meinkoth, J. and Wahl, G. (1984). *Anal. Biochem.*, **138**, 267.
5. Landegent, J. E., Jansen in de Wal, N., van Ommen, G-J. B., Baas, F., de Vijlder, J. J. M., van Duijn, P., *et al.* (1985). *Nature*, **317**, 175.
6. Harper, M. E., Ullrich, A., and Saunders, G. R. (1981). *Proc. Natl. Acad. Sci. USA*, **78**, 4458.
7. Furuta, Y., Shinohara, T., Sano, K., Meguro, M., and Nagashima, K. (1990). *J. Clin. Pathol.*, **43**, 806.
8. Lawrence, J. B., Villnave, C. A., and Singer, R. H. (1988). *Cell*, **52**, 51.
9. Harper, M. E. and Marselle, L. M. (1986). *Cancer Genet. Cytogenet.*, **19**, 73.
10. Cassab, G. I. and Varner, J. E. (1987). *J. Cell Biol.*, **105**, 2581.
11. Meinkoth, J. and Wahl, G. (1984). *Anal. Biochem.*, **138**, 267.
12. Rychlik, W. (1989). *Nucl. Acid Res.*, **17**, 8543.
13. Gottlieb, D. I. and Glaser, L. (1975). *Biochem. Biophys. Res. Commun.*, **63**, 815.
14. Hafen, E., Levine, M., Garber, R.L., and Gehring, W. J. (1983). *EMBO J.*, **2**, 617.

15. Maniatis, T., Fritsch, E. F., and Sambrook, J. (ed.) (1982). *Molecular Cloning. A Laboratory Manual*. Cold Spring Harbor Press, Cold Spring Harbor, NY.
16. Berger, S. L. and Kimmel, A. R. (ed.) (1988). *Guide to Molecular Cloning Techniques*. Academic Press, San Diego.
17. Cox, K. H., DeLeon, D. V., and Angerer, R. C. (1984). *Dev. Biol.*, **101**, 485.
18. Takahashi, T., Arakawa, H., Maeda, M., and Tsuji, A. (1989). *Nucl. Acid Res.*, **17**, 4899.
19. Wilchek, M. and Bayer, E. A. (1989). *TIBS*, **14**, 408.
20. Hopman, A. H. N., Wiegant, J., Tesser, G. I., and van Duijn, P. (1986). *Nucl. Acid Res.*, **14**, 6471.
21. Vernole, P. (1990). *BioTechniques*, **9**, 200.
22. Höltke, H-J. and Kessler, C. (1990). *Nucl. Acid Res.*, **18**, 5843.
23. Lion, T. and Haas, O. A. (1990). *Anal. Biochem.*, **188**, 335.
24. Melton, D. A., Kreig, P. A., Rebagliat, M. R., Maniatis, T., Zinn, K., and Green, M. R. (1984). *Nucl. Acid Res.*, **12**, 7035.
25. Cox, K. H., DeLeon, D. V., and Angerer, R. C. (1984). *Dev. Biol.*, **101**, 485.
26. Feinberg, A. B. and Vogelstein, B. (1983). *Anal. Biochem.*, **132**, 6.
27. Feinberg, A. B., and Vogelstein, B. (1984). *Anal. Biochem.*, **137**, 266.
28. Hodgson, C. P. and Fisk, R. Z. (1987). *Nucl. Acid Res.*, **15**, 6295.
29. Denhardt, D. (1966). *Biochem. Biophys. Res. Commun.*, **23**, 641.
30. Jackson, P., Dockley, D. A., Lewis, F. A., and Wells, M. (1990). *J. Clin. Pathol.*, **43**, 810.
31. Volkers, H. H., van de Brink, W. J., Leunissen, J. L. M., De Brabander, M., Zijlmans, H. J. M. A. A., Houthoff, H. J., *et al.* (1989). *Aurofile* **03**.
32. Angerer, L. and Angerer, A. C. (1989). *Du Pont Biotech. Update*, **4** (5).
33. Garson, J. A., van de Berghe, J. A., and Kemshead, J. T. (1987). *Nucl. Acid Res.*, **15**, 4761.
34. Zhang, F. R., Heilig, R., Thomas, G., and Aurias, A. (1990). *Chromosoma*, 99, 436.
35. McFadden, G. I., Bonig, I., Cornish, E. C., and Clarke, A. E. (1988). *Histochem. J.*, **20**,
36. Harris, N. and Wilkinson, D. G. (ed.) (1990). *In Situ Hybridisation: Application to Developmental Biology and Medicine*. Cambridge University Press.
37. Renz, M. and Kurz, C. (1984). *Nucl. Acid Res.*, **12**, 3435.

10

Problems and solutions

SUSAN VAN NOORDEN

1. Introduction

Successful immunolabelling, whether at light or electron microscopical level, is specific for a particular antigen, reliably repeatable, and gives a strong signal easily discernible against a negligible background. Beginners may be put off by less-than-ideal results, a peroxidase stain that can hardly be seen against the general background colour of the tissue section or gold particles scattered all over the ultrathin section with target localization not immediately obvious. However, the antigen–antibody reaction on which immunocyto-chemical labelling relies is extremely accurate and there is always a reason for a poor result. With experience and perseverence the cause of the disaster can usually be discovered and overcome, within the limits of the available reagents. This chapter will describe common, and some uncommon, problems in immunocytochemistry, along with remedies. Fuller discussion of all these points and others too specialized to be included here will be found in references (1–5).

The result of an immunocytochemical reaction is influenced by the amount of antigen present, the availability to applied antibody of its immunoreactive sites, the preservation of the tissue, the quality and specificity of the antibodies and other reagents (such as avidin) used, and the method. The problems, which may be connected with any or several of these factors, can be divided into three main categories:

- background problems—too much label
- method problems—too little label
- specificity problems—the 'wrong' result

These problems apply to any type of tissue sample for light microscopy or electron microscopy, whether frozen, paraffin, or resin sections, cells, cultures, or whole-mount preparations. Some problems fall into more than one category.

Of course, subjective judgement is used to decide whether there is a 'problem' when a new antibody, antigen, tissue, or method is being studied. It is therefore essential to carry through a minimum of two control preparations with every immunolabelling procedure. The first is a positive control preparation known to contain the antigen under investigation, to show that the reagents and method are all in working order. The second is a negative control preparation of the tissue being studied in which the primary antibody is omitted, or replaced by an inappropriate primary antibody, to show the level of background labelling that might be expected in that preparation with the method used, in addition to any due to the primary antibody.

2. Background problems

This heading covers the cases where the labelling of specific antigenic sites cannot be easily distinguished from labelling of the background in general or, in a few instances, particular tissue components such as connective tissue or blood cells, known not to contain the antigen in question. Unfortunately it is usually impossible to improve on a guess at the cause of the problem. Fortunately, however, it is not always necessary to know the exact cause since many common problems can be dealt with by similar treatment, diluting the primary antibodies and blocking unwanted binding sites. Once the problem has been conquered only the inquisitive with time on their hands will bother to investigate the precise underlying cause for it. Many of the possible causes of background labelling are given here together with remedies which should be tested for effectiveness in removing the problem.

2.1 Possible causes and remedies

2.1.1 Primary antibody too concentrated

This is the most common cause of trouble when a primary antibody is first used. Polyclonal antisera contain a mixed population of antibodies and other proteins that can bind to tissue non-specifically in various ways which are listed below. Provided the antibody is of high titre and avidity, the problem is easily overcome by dilution, and further causes and remedies need be sought only if this does not work. Monoclonal antibodies do not contain contaminants, unless they are presented in ascites fluid which contains proteins from the host animal, and should not, therefore, give unwanted staining, provided that suitable blocking of non-specific sites has been carried out (see 2.1.2 (a)). Usually about 10 µg/ml is a suitable concentration of antibody in a two step indirect method.

To overcome the problem dilute the primary antiserum further. Use a doubling dilution series on a known positive control prepared like your test tissue, with the method you intend to use and pre-established, standard dilutions of second and third reagents. Further refinement of dilution can be performed around the dilution that gives strong labelling with low background. A suitable dilution range to try for polyclonal antisera is 1/50–1/6400, for monoclonal antibodies in tissue-culture medium 1/10–1/1280, or in ascites fluid 1/100–1/12 800. Once a standard dilution has been found it is not usually necessary to alter it (*Figure 1*).

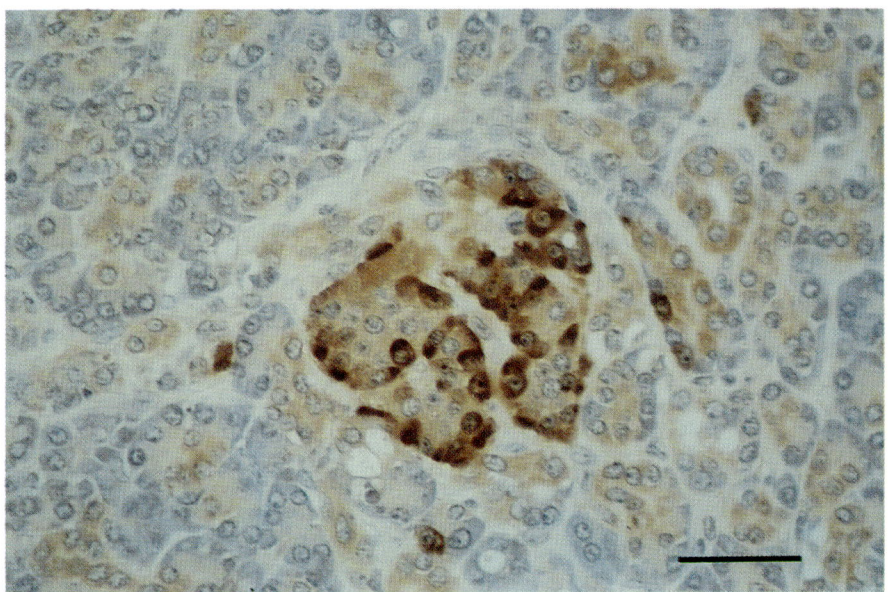

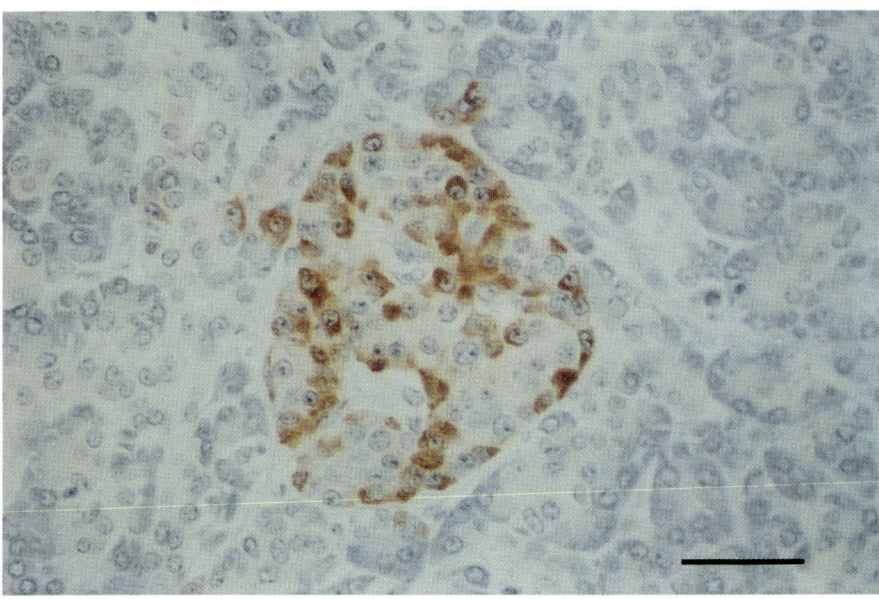

Figure 1. Formalin-fixed paraffin section of human pancreas immunostained for glucagon with a polyclonal rabbit antibody to glucagon applied overnight at 4 °C and the PAP method. Peroxidase developed with DAB and H_2O_2. In the top figure the primary antibody was used at a dilution of 1/100 and in the bottom figure at 1/2000. Note the reduction in background staining at the higher dilution. Scale bar 40 μm.

2.1.2 Charged tissue sites and hydrophobic bonding

Antibodies conjugated to a fluorescent label have an increased negative charge which attracts them to the positively charged acidophilic cytoplasm of tissue, especially in unfixed or lightly fixed preparations. Aldehyde fixation increases the negative charge of the tissue and reduces this danger, but increases the hydrophobicity of the tissue, encouraging hydrophobic bonding from all reagents. Both these types of non-specific bond are much weaker than the specific antigen–antibody reactions and can often be eliminated by dilution of the antibodies. Other remedies, listed below, are designed to block the non-specific binding sites or challenge weak bonds. It may be necessary to increase the concentration of the primary antiserum if these measures are adopted, as weak specific binding from low affinity antibodies in the population will be challenged along with the non-specific binding, and might reduce the overall binding of the primary antibody.

(a) Block the non-specific sites before applying the first antibody (*Figure 2*). Use non-immune (normal) serum from the species providing the second antibody (or primary antibody in a one step method), undiluted or up to 1/30 dilution for 10–30 min. Do not wash off, merely drain before applying the first antibody. Inert protein such as bovine serum albumin (BSA), gelatin, or chicken egg albumin can be used instead at a dilution of 1% (w/v).

  i. Use serum as a blocker on fresh tissue to block Fc receptors (see 2.1.3).

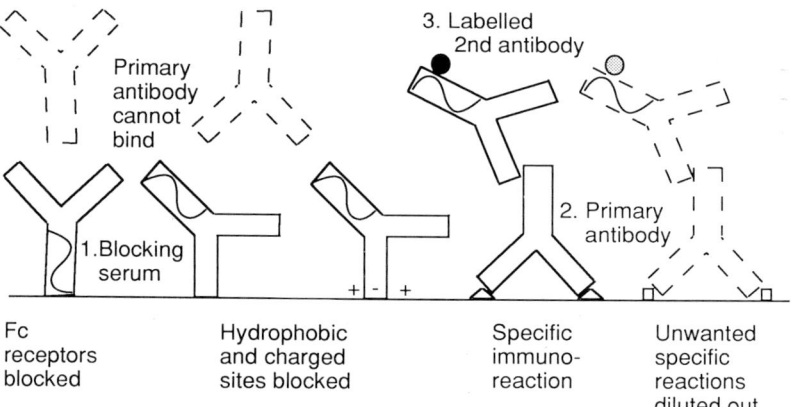

| Fc receptors blocked | Hydrophobic and charged sites blocked | Specific immuno-reaction | Unwanted specific reactions diluted out |

Figure 2. Non-specific binding of primary antibody is prevented by prior application of normal serum from the species in which the second antibody is raised. 1, Blocking layer: immunoglobin in normal serum from the species providing the second antibody blocks non-specific binding sites. 2, First antibody: dilute primary antibody finds specific antigen; non-specific sites blocked; minority unwanted antibodies diluted out. 3, Second antibody: labelled second antibody binds to primary antibody, but not to immuno-globulins of the same species blocking non-specific sites. Unwanted reactions are negligible.

  ii. Do not use serum as a blocker for Protein A methods, since Protein A binds to immunoglobulins. Use an inert protein.

  iii. Do not use chicken egg albumin in avidin–biotin methods as it contains avidin.

  iv. Do not use BSA if this protein was used as a carrier protein in the production of the primary antiserum, which may therefore contain a population of antibodies to BSA.

  v. As a precaution against weak blocking proteins being washed off the preparation, repeat the blocking step before applying the second antibody.

(b) Add detergent to buffer and/or antibody diluent. Suitable additives are 0.2% Triton X-100 or 0.1% saponin, which is less likely to extract membrane proteins (6). **Take care** as reduced surface tension may cause solutions to spread over a slide. Encircle the section with water-repellant such as with a wax pencil or a proprietary hydrophobic pen (Dako, High Wycombe, UK; Miles Labs, Slough, UK) or simply rinse in buffer plus detergent after the primary antibody, followed by buffer without detergent and finally dry around the section.

(c) Increase the sodium chloride concentration in the antibody diluent and/or buffer to 2.5% (7).

(d) Increase the pH of the buffer to 9 or 10 (7), particularly for avidin which has a pI of 10, since this inhibits its binding to charged tissue sites (8).

2.1.3 Fc receptors

Fc receptors, found on macrophages and some T lymphocytes, bind immunoglobulins non-specifically, usually weakly but particularly strongly with complexes such as peroxidase–anti-peroxidase that can attach to two receptors at once. This is only a problem in fresh tissue. Fixation and processing destroys the receptors.

(a) Perform a preliminary block with immunoglobin or serum from the species providing the second antibody (or primary antibody in a one step method) as 2.1.2 (a).

(b) Use F(ab′)$_2$ fractions of antibodies (with the Fc portion removed by digestion with pepsin). It may be necessary to use antibodies to the F(ab′)$_2$ fraction in the subsequent step.

2.1.4 Complement fixation

Serum C1q, the first component in the complement cascade, may bind to receptors in the tissue and precipitate immunoglobulin complexes from the serum by attachment to the complement-binding areas of the Fc portions of IgG and IgM. This probably only occurs with certain tissues and low dilutions of antisera and does not seem to be a common problem.

(a) Decomplement all sera by heating to 56 °C for 30 min in a water bath (there seems to be some doubt as to whether C1q is inactivated by this procedure).

(b) Absorb C1q from antisera by adding anti-C1q or absorbing against an irrelevant antigen–antibody complex (9).

(c) Use F(ab')$_2$ fractions of antibodies.

2.1.5 Cross reacting immunoglobulins

The second antibody contains a population of immunoglobulins that cross react (specifically) with host tissue immunoglobulins, causing unwanted labelling of plasma cells, interstitial areas, and other immunoglobulin-rich sites in the host tissue.

(a) Absorb cross reactivity from the diluted second antibody by adding to it 1–2% (v/v) of serum from the host species. Check by using the second antibody with and without absorption to label immunoglobulins in spleen of host tissue, prepared in same way as your test preparation or as acetone-fixed cryostat sections.

(b) Use a commercial species-specific anti-immunoglobulin.

(c) Make a complex of the secondary labelled antibody with the primary (10). Find the optimal proportions of primary and labelled secondary antibodies by a checkerboard series of mixtures at different dilutions. Eliminate any remaining free anti-primary species immunoglobulin binding sites by adding 1% of normal primary species serum after the complex has formed (1 h is adequate). This ingenious method will even allow use of a primary antibody raised in the same species as the host tissue. For example, a mouse monoclonal antibody to an antigen in mouse tissue can be complexed with peroxidase-conjugated anti-mouse immunoglobulin raised in any species and used to label mouse tissue.

2.1.6 Binding of proteins by amino acids

Basic amino acids in tissue proteins can bind immunoglobulins and other proteins such as streptavidin and Protein A, particularly when they are aggregated in solution due to labelling (11). This gives rise to labelling and absorption controls that are apparently specific and is discussed further in Section 4.1.3. To correct this absorb non-specific reactivity of the antibody by adding to the diluted antibody a basic peptide: poly-L-lysine, mol. wt. 3000–6000, 2 mg/ml (11). This does not affect specific antigen–antibody reactions and could be added to all immune reagents as a precaution.

2.1.7 Binding of carrier protein antibody

Antibody to carrier protein used in immunization binds (specifically) to carrier protein in tissue (such as albumin). To overcome this add 1% (w/v) carrier protein to the diluted antibody.

2.1.8 Biotin-rich tissue

Tissues rich in biotin, such as liver and kidney, bind labelled avidin or streptavidin complexes. In these cases pre-treat the preparation with unconjugated avidin (1 mg/ml for 20 min) to block the avidin-binding sites, rinse in buffer and apply unconjugated biotin (0.1 mg/ml for 20 min) to block any free biotin-binding sites on the applied avidin. Rinse again and continue with the immunoreaction. Since biotin molecules have only one avidin-binding site, no further avidin will be attracted (12).

2.1.9 Binding by labelled avidin

Labelled avidin binds to mast cells (heparin) and other charged sites in the tissue because of its high pI.

(a) Raise the pH of the buffer used to dilute the avidin to more than 9. Tris/HCl buffer, pH 9.2 or 0.1 M carbonate buffer, pH 9.4 can be used (8).

(b) Block tissue sites with a basic peptide, poly-L-lysine, mol. wt. 3000–6000, 0.01% (w/v) in PBS, pH 7.6 for 30 min, (8).

(c) Use a 'modified avidin' (commercially available) or streptavidin reagent (neutral pI) instead of an avidin reagent.

2.1.10 Binding of avidin to lectins

Avidin binds to lectins in tissues through its carbohydrate groups (lectins bind to specific sugars).

(a) Add 0.1 M α-methyl-D-mannoside to the avidin solution (13).

(b) Use streptavidin (protein, not glycoprotein) instead of avidin.

2.1.11 Endogenous peroxidase

Endogenous peroxidase occurring in macrophages, other blood cells, and kidney tubules is demonstrated together with the applied immunoreagent-linked peroxidase. This is mainly a problem with fresh tissue as activity is destroyed by processing to paraffin except in red blood cells (pseudo-peroxidase).

Endogenous peroxidase activity can be inhibited with excess of its substrate, hydrogen peroxide, before beginning the immunoreaction. If the antigen to be localized is destroyed by this treatment, perform it after the primary antibody step but before applying the peroxidase-linked reagent. The hydrogen peroxide concentrate used must be within its shelf-life period. Formulae for peroxidase blocking solutions are given below.

(a) For fresh cryostat sections or cell preparations (gentle methods): 0.1% phenylhydrazine (freshly prepared) in PBS at pH 7.0 for 1 h at 37 °C (14).

(b) Azide, itself an inhibitor of peroxidase, in combination with nascent H_2O_2 (*Protocol 1*), which is produced from the action of glucose oxidase on glucose (15) as follows:

$$\text{glucose oxidase}$$
$$\beta\text{-D-glucose} + O_2 + H_2O = \text{D-gluconic acid} + H_2O_2$$

Protocol 1. Mild blocking method for endogenous peroxidase using nascent H_2O_2 and sodium azide

1. Pre-warm 100 ml PBS to 37 °C.

2. Just before use add:

- 1 ml of 18% glucose (gives 10 mM)
- 1 ml of 0.65% sodium azide (gives 1 mM)
- glucose oxidase to give 1 U/ml

3. Incubate acetone-fixed cryostat sections or other preparations at 37 °C for 1 h.

(c) For paraffin or resin sections: 3% H_2O_2 in water, 0.01 M phosphate-buffered 0.9% sodium chloride, pH 7.0–7.4 (PBS) for 5 min, or 0.3% H_2O_2 in methanol, or 70% methanol in PBS for 30 min at room temperature.

(d) For paraffin sections with intransigent peroxidase activity from a high content of red blood cells see *Protocol 2*: **care**—this method may damage some antigens such as leucocyte common antigen.

Protocol 2. Strong blocking method for endogenous peroxidase (16)

1. Bleach acid haematein with 6% H_2O_2 (6 ml of 30% H_2O_2 in 100 ml distilled water) for 5 min at room temperature. Rinse in tap water.

2. Inhibit peroxidase with 2.5% periodic acid in distilled water for 5 min at room temperature. Rinse in tap water.

3. Block any free aldehyde groups created by periodic acid oxidation or left in tissue by aldehyde (see Section 2.1.14) with freshly made 0.02% sodium or potassium borohydride in distilled water for 2 min at room temperature. Rinse in water, then buffer.

2.1.12 Peroxidase-linked immunoreagents

Peroxidase-linked immunoreagents in solutions containing calcium ions can bind via the enzyme to mannose-specific lectin-like sites in the tissue (17). This does not seem to be a common problem.

(a) Avoid calcium-containing buffers.

(b) Block tissue sites with mannose.

(c) Apply peroxidase (0.15% (w/v) in PBS) to the tissue before inactivating endogenous peroxidase. The applied peroxidase will block the binding sites and be inactivated together with the endogenous enzyme. This method is suitable in a PAP reaction but cannot be used if the detection sequence contains an unbound anti-peroxidase (18).

2.1.13 Endogenous alkaline phosphatase

Endogenous alkaline phosphatase (in fresh tissue only) is reactive. Include 10 mM levamisole in the final enzyme-incubating medium. This inhibits all alkaline phosphatase isoenzymes except the intestinal alkaline phosphatase linked to the immune reagent (19). Alkaline phosphatase is an unsuitable label for immunocytochemistry on intestinal tissue.

2.1.14 Aldehydes remaining in tissue

Aldehydes left in the tissue by fixatives (formaldehyde and particularly glutaraldehyde) provide covalent binding sites for applied immunoreagents via their amino groups. If not sufficiently blocked by serum alone, the following can be tried:

(a) 0.02–1% sodium or potassium borohydride in 0.1 M phosphate buffer, pH 7.4 (20) or water for ultrathin, epoxy resin sections (21) for 30 min at room temperature

(b) 50–100 mM ammonium chloride added to the blocking serum

(c) 100 mM ethanolamine added to the blocking serum

(d) 0.2 M glycine in PBS for 5 min

2.1.15 Pigment in tissue

Pigment in tissue gives the impression of a positive DAB–immunoperoxidase or silver-intensified reaction.

(a) Remove pigment at a suitable stage:

    i. Mercury pigment (from fixatives containing mercuric chloride) can be removed by standard Lugol's Iodine and sodium thiosulphate treatment before performing the immunoreaction or after the peroxidase has been developed.

    ii. Melanin can be bleached before performing the immunoreaction.

    iii. Formalin pigment can be removed by soaking sections for 15 min to several hours in equal parts of absolute alcohol and picric acid saturated in alcohol before or, preferably, after performing the immunostain (22).

(b) Use a different chromogen or enzyme label to give a contrasting colour.

2.1.16 Silver deposition

Silver from enhancement solutions is deposited on the tissue. To prevent this dip the slide carrying the sections in 0.5% gelatin and allow it to dry before silver intensification to protect the preparation. Afterwards, soak in warm water to remove the gelatin (23).

2.1.17 Staining of argyrophilic tissue components

Argyrophilic tissue components stain in silver development solutions.

(a) Keep the development period as short as possible. A short development period for enhancing a peroxidase reaction is facilitated by using a method that gives an intense reaction initially, e.g. with nickel sulphate/ DAB/H_2O_2 for peroxidase.

(b) Try various methods for suppressing tissue argyrophilia such as treatment with thioglycolic acid (24) or lanthanum nitrate and hydrogen peroxide (25). These are performed after the DAB development step. One of the simplest is given in *Protocol 3*.

Protocol 3. Suppression of tissue argyrophilia by copper sulphate (25)

After the DAB development:

1. Wash twice, 1 min each, in 1% (v/v) acetic acid.
2. Immerse for 10 min in 0.01 M copper sulphate.
3. Wash twice, 1 min each, in distilled water.
4. Immerse in 1% sodium acetate trihydrate containing 3% (w/v) hydrogen peroxide.
5. Wash twice, 1 min each, in 1% sodium acetate.
6. Continue with silver intensification.

2.1.18 Autofluorescence of tissue components

Tissue components show autofluorescence which interferes with applied immunofluorescence. Autofluorescence is usually greenish-yellow with the filters used to examine fluorescein fluorescence and may still show through the green filter used for rhodamine fluorescence.

(a) Mask the autofluorescence with a fluorescent background dye. Pontamine Sky Blue (0.05% (w/v) in PBS with 1% (v/v) dimethyl sulphoxide) is a useful substance, fluorescing red with the filters used for viewing fluorescein fluorescence. Stain the preparations for 30 min before applying the primary antibody (modified from 26).

(b) Use an alternative fluorescent label, such as rhodamine or Texas Red instead of fluorescein (red fluorescence) or 7-amino-4-methylcoumarin-3-acetic acid (AMCA) (blue fluorescence) (27).

2.1.19 Deposition of aggregates
Aggregates in gold-labelled antibodies are deposited on ultrathin sections on the grid. To prevent this centrifuge gold-adsorbed antibodies at 2000g for 10 min and use the supernatant.

2.1.20 Labelling of trapped antibody
Antibody trapped in tissue or dried on the preparation is labelled. To prevent this take care to use unwrinkled sections, wash very thoroughly at all stages, and avoid drying the preparations during the procedure.

2.1.21 Attraction of reagents by slide
Reagents are attracted to the positively charged slide when poly-L-lysine is used as a slide coating to ensure section adhesion (28). The deposit is not usually on the tissue itself, since this overlies the coating, but around the section and on exposed slide where there are spaces in the tissue. Thus it does not necessarily interfere with the reading of the result but may be unsightly.

(a) Make sure the poly-L-lysine is spread evenly and thinly on the slides.
(b) Try an alternative slide coating such as silane, used for slides for *in situ* hybridization of DNA or mRNA (29), Vectabond (Vector Laboratories, Peterborough UK, USA), or no slide coating, simply very clean slides.

2.1.22 The peroxidase reaction
The peroxidase reaction is compromised by traces of bleach (sodium hypochlorite, sometimes used to decontaminate DAB solutions) remaining on the walls of the DAB reaction vessel, causing partial oxidation and coloration of the DAB solution during the reaction: glassware used for DAB solutions must be washed very thoroughly after decontaminating.

2.1.23 Leaching of immunoreactive substances
Immunoreactive substances leach from necrotic cells into the surrounding tissue or necrotic cells in the tissue and take up antibody. There is no remedy to this problem; simply ignore the necrotic area.

2.1.24 Phagocytosed antigen
Phagocytosed antigen gives specific localization, but in the 'wrong' place. Again, there is no remedy for this.

2.1.25 Diffusion of reaction product
Polymerised DAB reaction product can creep along membranes giving an inaccurate impression of antigen localization. This can be a particular problem in pre-embedding immunoperoxidase reactions for electron microscopy in thick specimens when a long incubation time has to be used.

(a) Keep the incubation period in DAB solution short.

(b) Use another method.

3. Method problems

Finding the cause of too little immunolabelling is a trickier problem. Much will depend on whether you have confidence in the antibodies you are using. In this section it is assumed that the detecting antibodies and other reagents are tried and trusted, and that it is the primary antibody or the tissue localization of the antigen that is unknown.

It cannot be stressed too strongly that it is vital to include a positive control preparation for each primary antibody in each staining run, preferably prepared in the same way as the test tissue. Only if your control is positive and up to its usual standard can you be certain that you have performed the method correctly and that all stages are in good order. The purist will include a weakly staining control tissue as well as a strongly staining one so that a decline in the quality of the reagents is noticed from the poor staining of the weak control, even if there is still enough staining in the strong positive to mask the decline. In a routine diagnostic laboratory where repeated staining for a variety of standard antigens is required, a composite block containing several different tissues is a useful device for reducing both the number of control blocks to be cut and the likelihood of error in choosing a control preparation.

3.1 Common errors

3.1.1 Human error

This is by far the most common cause of failure in day-to-day immunostaining—taking the wrong bottle from the refrigerator, making an error in calculating a dilution, not checking a pH, forgetting the hydrogen peroxide in the peroxidase developing reagent, adding the 'blocking' concentration of hydrogen peroxide instead of the substrate concentration, making up a peroxidase-linked reagent in buffer containing azide which inhibits the enzyme activity—all these mistakes have occurred often in my laboratory, usually when the experimenter is a beginner; the list could be endless. To discover the mistake retrace your actions: if both your control and test staining have failed and you cannot find a fault along these lines repeat the method as before with new preparations. If it fails again, it is time to check the reagents themselves.

3.1.2 Reagent fault

Immune reagents can deteriorate from several causes described in more detail below. At this stage the main point is to identify which of the reagents is causing the problem. Substitute a different positive primary antibody to check the rest of the reagents. If staining is positive on the second positive control, your original primary antibody must be at fault. If staining is negative, the fault probably lies with one of the detecting reagents which must be separately checked.

3.1.3 Saving an immunostain

It may be important to produce a result for urgent diagnosis, and even if something has gone wrong with one of the steps of the method, adding further layers to build up the reaction may allow a result to be read, even if the labelling is not optimal. The failed positive control must of course be included in the saving operation. Although the following paragraphs refer to peroxidase methods, they are just as relevant to other labels.

(a) An indirect two step method may be saved by reapplying the primary, provided that an excess of second antibody was originally applied so that some anti-primary immunoglobulin sites are still unoccupied. This is followed by reapplication of the second antibody and redevelopment (*Figure 3*).

(b) A PAP stain may be saved by repeating the second and third layers (with freshly made dilutions) and the peroxidase development, assuming the primary antibody is correct.

   i. If the original second layer did not bind to the primary or was too dilute, the repeat application should find the unoccupied sites.

   ii. If the PAP reagent failed on the first application it will now have another chance.

   iii. If all the immune reagents were well bound at the first attempt but the peroxidase reaction was blocked or damaged, the second

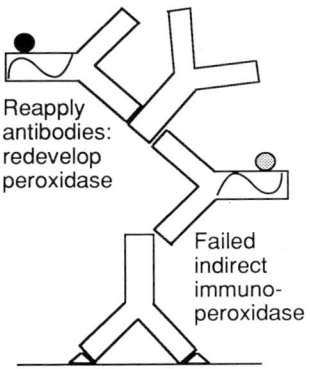

Reapply antibodies: redevelop peroxidase

Failed indirect immuno-peroxidase

Figure 3. Saving a two step indirect immunoperoxidase stain by reapplication of primary and secondary antibodies. If the first two layer method was unsuccessful through failure of enzyme development, reapplied primary antibody will bind to free sites on second layer labelled antibody of the first reaction. Reapplied second labelled antibody will then bind to the newly applied primary.

application of the anti-immunoglobulin will bind to the PAP (immunoglobulin—peroxidase complex) and the second PAP will finish the reaction ready for redevelopment (*Figure 4*).

(c) An ABC stain may be saved by applying a layer of appropriate PAP and redeveloping, provided that the concentration of the second (biotinylated) antibody was high enough to leave unoccupied anti-IgG sites free (*Figure 5*). It is doubtful whether repeating the biotinylated antibody would find any further unoccupied primary antibody, unless the wrong anti-species biotinylated antibody has been used in the first attempt, and repeating the ABC will not help because all available biotin sites on the second antibody should have been occupied by avidin already.

(d) If an alcohol-soluble enzyme reaction product is removed by inadvertently dehydrating the stained section through alcohols, the preparation can be saved by repeating the steps of the reaction as in (a)–(c) above and redeveloping. Mount in a water-based mountant.

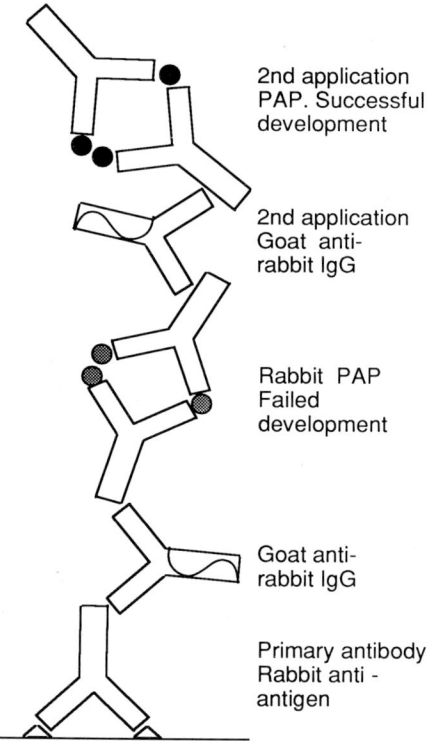

2nd application PAP. Successful development

2nd application Goat anti-rabbit IgG

Rabbit PAP Failed development

Goat anti-rabbit IgG

Primary antibody Rabbit anti-antigen

Figure 4. Saving a PAP stain by reapplication of the second and third immunoreagents. In a failed PAP stain, assuming the immunoreagents were successfully applied, a repeated application of the second antibody will find immunoglobulin-binding sites on the original PAP complex. A second application of PAP will bind to the newly applied second antibody.

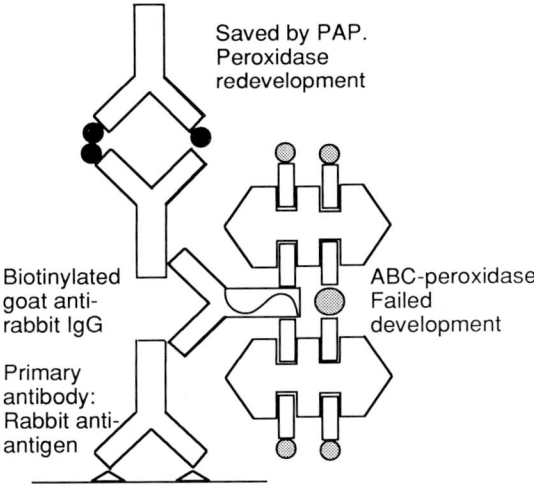

Figure 5. Saving an ABC-peroxidase stain by application of a PAP step. Assuming the second (biotinylated) antibody and ABC have bound, but development failed, the simplest rescue strategy is to apply PAP (same species as primary antibody) to bind to free anti-immunoglobulin sites of second antibody.

(e) A faded indirect immunofluorescence preparation can be revived by repeating the fluorescent conjugate step (30). The only logical basis for this finding is that the original fluorescent antibody has become detached from the primary, but this has not been proved. Success of this stratagem, even when fluorescence has faded during microscopic viewing, has been reported (reference 5, p. 82).

3.2 Method factors affecting immunolabelling

Section 3.1 has dealt with common errors and how to overcome them in a particular manner. In the following section the many factors that can prevent optimal immunostaining will be discussed.

3.2.1 Sample preparation

If the test tissue and the positive control tissue have been fixed and processed in the same way, and the control is adequately labelled, a negative result on the test tissue must be accepted as genuine. If identical conditions do not apply, it may be necessary to look for a remedy for an unexpected negative result.

i. Over-fixation

Although fixation is usually necessary to make antigens insoluble and preserve the structure of the tissue, it can alter antigenic sites so that they are no longer recognized by the antibody which was raised to a pure, unfixed antigen. Over-fixation with aldehydes can cross-link protein and peptide molecules so that the epitopes are not available to the applied antibody. Monoclonal antibodies, in particular, may be severely affected by fixation, since each reacts with only one epitope. Polyclonal antibodies have a better chance of finding undamaged epitopes. If a monoclonal antibody gives a negative result in tissue fixed with glutaraldehyde for ultrastructural studies, it is worth trying formaldehyde instead, since this is a weaker cross-linking fixative that can still give adequate ultrastructural preservation. Some antibodies will only react with their antigen in frozen sections, lightly post-fixed in a precipitating fixative such as acetone. Frozen ultrathin sections (pre-fixed) or freeze-dried tissue, post-fixed in osmium or other fixative vapour, are also possibilities for electron microscopical immunolabelling (4, Chapter 16). If over-fixation is suspected:

(a) Try other fixatives.

(b) Try a shorter fixation time.

(c) Try washing the fixative out of the sections before staining or out of the block before processing. This can be effective for formaldehyde fixation which is partially reversible.

(d) Treat aldehyde-fixed paraffin sections with protease before beginning the immunolabelling procedure (31) (see *Protocol 4*). This is very effective for large molecules, and is thought to 'undo' the cross-links formed during aldehyde fixation. Trypsin (*Protocol 4*) is cheap and generally useful but it may be necessary to try several different enzymes to select the best for a particular antigen (32–34). The temperature and pH of the protease solution must be optimal for the particular enzyme and a standard way of treating the preparations must be used. If one person in the laboratory pre-warms the slides while another starts the process with cold slides, yet both use the same time in the enzyme solution, different results will be obtained. It is best to establish a standard time for each antigen in the type of preparation you are using, as some require much longer treatment than others. The activity of trypsin (and other enzymes) is greatest at the beginning of the incubation period so preparation of the solutions should be done as quickly as possible. Standardization is easiest and protease action most uniform if sections are incubated by immersion in a large volume of the protease solution. However, for the more expensive enzymes the advantage is outweighed by the cost, and in this case, standardization should be done with drops of enzyme solution on the preparation. If timing seems to be critical, staggered application will be necessary when slides are being treated in large numbers. Conditions must be tested each time a new batch of enzyme is used. It is advisable to have paraffin sections mounted on slides coated with poly-L-lysine (28) or other adhesive to prevent detachment of sections. Enzyme treatment may be performed before or after hydrogen peroxide blocking of endogenous peroxidase.

Protocol 4. Trypsin treatment to reveal aldehyde cross-linked antigens

Care: highly purified trypsin may not be useful. A suitable one (catalogue no. 150213) is obtainable from ICN-FLOW, High Wycombe, UK.

1. Pre-warm 0.1% (w/v) calcium chloride to 37 °C (you can keep a large volume in a bottle ready in a 37 °C incubator).
2. Check that pH meter is registering correctly.
3. Add 0.1 g trypsin to 100 ml pre-warmed 0.1% (w/v) calcium chloride. It will not dissolve completely at this stage. The pH will be between 5 and 6.
4. Bring the pH of the solution to 7.8 with 0.1 M sodium hydroxide. Do this as quickly as possible to avoid cooling the solution. Most of the trypsin will dissolve. Immerse the sections in the solution.
5. Incubate sections for the established time at 37 °C. The incubation period may vary between 5 min and 75 min but is usually 10–15 min. Semi-thin resin sections (resin removed) may need as little as 30 sec.
6. Stop the incubation by rinsing the sections well in cold water.

- For protease treatment, make a small volume of 0.05% (w/v) protease (e.g. Protease XXIV, P8038, Sigma, UK, USA) in pre-warmed PBS. There is no need to check the pH which does not alter. Use the solution as drops on the sections for 5–20 min at 37 °C.
- For pepsin treatment, use pepsin (ED 3.4.23.1, Sigma P7012, Sigma, Poole UK) at 0.4% (w/v) in 0.01 M HCl at 37 °C for 20–30 min.

(e) Treat osmicated ultrathin resin sections on grid with saturated (5% w/v) aqueous sodium metaperiodate to reverse the effect of osmium fixation (35). This is effective for some antigens. The optimum period must be tested for each preparation.

ii. Under-fixation

Some antigens such as some neuropeptides may be so soluble that in fresh frozen tissue they diffuse from their tissue site or dissolve as a cryostat section is cut. Diffusion can also be a problem in sections from under-fixed paraffin blocks. Diffusion and relocation of antigen may be even more important at the sub-cellular level than for light microscopy, but is hard to prove. If immunolabelling gives unexpected results, try a variety of fixatives, including freeze-drying as in (b) below (reference 4, Chapter 16).

(a) Try fixing the tissue block before freezing (for antigens destroyed by processing but not by fixation). Before freezing, the fixed block should be soaked in a cryoprotectant such as gum sucrose (dissolve 1 g gum acacia (gum arabic) mixed with 30 g sucrose in 100 ml water with gentle heat and stirring; cool to 4 °C before use) or PBS containing 15% (w/v) sucrose.

(b) Try freeze-drying the tissue followed by vapour-fixing and embedding in paraffin or resin under vacuum (4, Chapter 16, 36, 37). This method requires special equipment.

iii. Antigenic sites obscured in tissue-bound antigen

If the antibody was tested *in vitro* by enzyme-linked immunosorbent assay (ELISA) or radioimmunoassay (RIA) it may react with epitopes that are not available on the fixed antigen *in situ*. Always test antibodies in the system in which they are to be used, such as by immunocytochemistry using tissue prepared in the way in which you intend to use it. Do not rely on RIA, ELISA, or immunodiffusion tests—they may help to define specificity but will not tell you if the antibody can be used immunocytochemically. Check this with the supplier of your antibody before purchase.

iv. Processing

Only drastic errors will result in complete lack of labelling, since antigens should be properly fixed before processing through alcohols to wax or resins. Some antigens may be damaged by heat arising through polymerization of resin in embedding for electron microscopy. Some may be washed out of tissue by alcohols or by long storage in buffer which can reverse the fixation effect of formaldehyde.

(a) Keep the processing time as short as possible but consistent with adequate dehydration and infiltration of the embedding medium.

(b) Use wax of low melting point (45 °C). This may, however, be difficult to cut.

(c) Embed and polymerize in epoxy resins at low temperature (37 °C).

(d) Use ultraviolet light to polymerize epoxy or acrylic resins at 4 °C. Polymerization time will be longer than the usual heat-assisted time for both these procedures.

(e) Use a special low temperature embedding method and acrylic resin (Lowicryl K4M or HM20).

(f) Try frozen sections instead of paraffin or (for electron microscopy) resin sections.

v. Sectioning

Soluble and poorly fixed antigens may diffuse in sections (see Section 3.2.1.ii).

Some antigenic sites may be destroyed by using a hot plate to ensure adhesion of paraffin sections to slides. **Do not** use a hot plate at a temperature greater than 45 °C. Mount paraffin or resin sections on coated slides (see Section 2.1.21) and dry for at least 3 h (better overnight) at 37 °C.

vi. *Physical masking of antigens*
Immunolabelling of resin sections occurs on the surface of the section only. If the antigen sought is on a structure smaller than the thickness of the section, such as viruses, some of the sites may not be exposed to the reagents (38). Immunolabelling of many serial thin sections should expose the antigen in at least some of its sites.

vii. *Hydrating wax or resin sections (for light microscopy)*
Removal of wax or resin (for light microscopy) from sections must be complete to ensure that antibodies can reach the preparation. Take care with removal of wax or resin.

viii. *Permeabilization*
It may be necessary to permeabilize whole-cell or whole-mount preparations or thick, cryostat, or vibratome sections with detergent if fixation did not dissolve lipids from membranes and intracellular antigens are to be detected.

(a) Try treatment for 30 min at room temperature with a variety of detergents dissolved in buffer before beginning the immunolabelling procedure. Suitable solutions are Triton X-100 (0.05–1%), Tween 20 (0.05–2%), or Saponin (0.05–0.1%) which is less likely to extract membrane proteins (6). The concentrations given here are only a rough guide. Try a range.

(b) Penetration of reagents into pre-fixed unprocessed thick sections and whole mounts is often helped by dehydrating the preparation through a series of graded alcohols to a solvent then rehydrating (39).

(c) Ultrathin, epoxy resin sections on grid may be 'etched' with 10% hydrogen peroxide for up to 10 min before immunolabelling. Take care with this treatment as it may damage the tissue.

(d) Some epoxy resin may be removed from ultrathin sections on grid by brief treatment with alcoholic sodium hydroxide (40) in a much weaker solution than used for semi-thin sections on slides (41).

ix. *Blocking endogenous peroxidase*
Hydrogen peroxide blocking (see Section 2.1.11 above) can damage some antigens, particularly cell surface antigens on frozen sections or in whole cell preparations.

(a) Use a milder blocking method:
- 0.3% hydrogen peroxide in 70% methanol in buffer for 30 min
- nascent hydrogen peroxide with azide (15) (see *Protocol 1*)
- phenylhydrazine 0.1%, pH 7.0. 37 °C for 1 h (14)

(b) Perform the peroxidase blocking step after application of the primary antibody but before application of the peroxidase blocking reagent.

(c) Use a different enzyme label.

x. *Buffer*

The pH is important. Too low or too high a pH may cause dissociation of antibodies from their antigens, particularly if the bonds are of low affinity: check the pH of all buffers (pH 7.0–7.6 is a safe range).

xi. *Temperature*

Immunoreactions, like other physico-chemical reactions, proceed more slowly at low than at high temperatures. A lower than usual temperature in the laboratory may reduce the rate of the reaction: check the temperature of the solutions.

xii. *Timing*

Adequate time must be allowed for the antibody to reach equilibrium with the antigen. The more dilute the antibody solution (high dilutions are preferred for reducing contaminants and for economy) the longer the time required. Incubate with a highly diluted polyclonal primary antibody for at least 4 h. Overnight incubation (at 4 °C to prevent evaporation from the drops of antibody solution on the slides) may be convenient and even longer (48 h) may improve the reaction as the high affinity antibody population gradually replaces the low affinity antibodies that find some of the antigenic sites.

- A shorter incubation time is preferable for fresh (lightly post-fixed) frozen sections or whole cells, to avoid damage to the tissue.

- Monoclonal antibodies are of uniform specificity and affinity and therefore do not need to be diluted so highly. Shorter incubation times (1–2 h at room temperature) are therefore satisfactory.

xiii. *Sensitivity*

The method may not be sufficiently sensitive to detect the small amount of antigen present in the preparation.

(a) Try a more sensitive method (a three step method instead of a two step method, or an ABC method instead of a PAP method).

(b) Try giving greater visibility by enhancing the intensity of the reaction product (silver intensification of immunogold (42–44) or peroxidase/DAB (45, 46).

(c) For immunogold labelling at electron microscopical level, the smallest sized gold particles (1–5 nm diameter) are most efficient, but they are difficult to see at low magnifications. Silver (47) or gold (48) intensification can help here.

3.2.2 Antibody factors

i. Dilution

A standard dilution should be established on a positive control but if a different preparation method is used or very little antigen is present the dilution may be too high. Increase the concentration of the primary antibody to give a greater proportion of high affinity antibodies in the diluted sample. Monoclonal antibodies, of uniform affinity, would not be affected.

ii. Bigbee (prozone) effect

In an enzyme–anti-enzyme method, the second antibody forms an immuno-globulin bridge between the first and third reagents via its two binding sites. If it is too dilute, both binding sites will be occupied by the primary antibody, thus preventing bridge formation. The second antibody must, therefore, always be applied in excess with respect to the primary (49) (*Figure 6*).

(a) Increase the concentration of the second antibody.

(b) Increase the dilution of the primary antibody.

iii. Affinity

A polyclonal antiserum may contain low and high affinity antibodies. Some antigenic sites will be occupied initially by low affinity antibodies which are washed off during the various washing steps of the procedure, reducing the

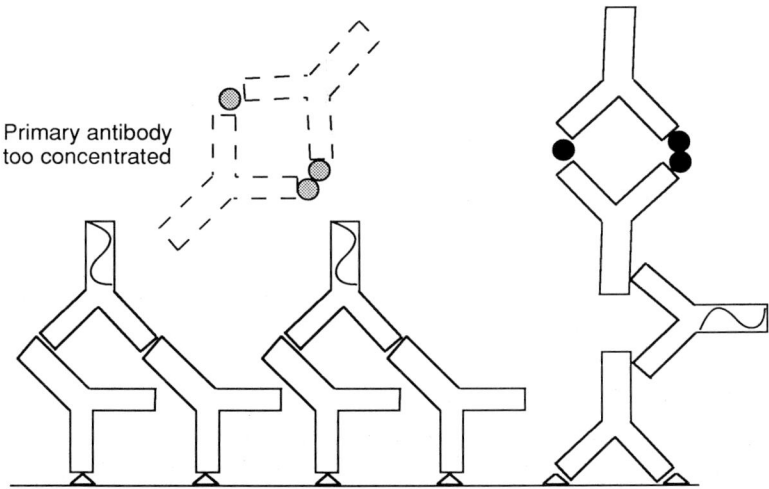

Primary antibody too concentrated

Figure 6. The Bigbee (prozone) effect. When the primary antibody is highly concentrated with respect to the second antibody, both immunoglobulin-binding sites of all applied second antibody molecules will be linked to the primary antibody, and none will be available to bind to the PAP reagent (left side of figure). An excess of second antibody over primary antibody is needed for success with the PAP method (right side of figure).

amount of final label. Half-way through the incubation period with the primary antibody rinse the preparations to wash off low affinity antibodies and reapply the primary antibody at the same dilution as before to allow the high affinity population to bind to the freed antigenic sites (50).

iv. Protein A

Protein A has variable affinity for immunoglobulins of different species. If it is suspected that a Protein A–gold reagent is not binding to the antibody species of the preceding layer substitute gold-adsorbed Protein G, or a new recombinant reagent consisting of a mixture of Proteins A and G (51).

v. Steric hindrance

In concentrated solutions large molecules such as avidin-biotinylated label complexes or antibodies adsorbed to colloidal gold particles of large diameter may be prevented from reaching their target antigens. Use a series of further dilutions to establish the optimum concentration.

3.2.3 Storage

It is important to store all immune reagents in suitable conditions. Guidelines are usually given by the commercial supplier, but the main points are that enough reagents for current use should be kept at 4 °C, with added protein (see ii below), and the stock kept snap-frozen in aliquots (provided it withstands freezing). Sodium azide (0.01–0.1%) should be added to reagents kept at 4 °C. Although azide inhibits peroxidase activity and should not be added to diluted peroxidase-linked reagents, some suppliers add it to the concentrated reagent. After dilution for use, the concentration of azide will be too low to inhibit the enzyme. Merthiolate is an alternative preservative. Even following these recommendations, problems may occur.

i. Antibody destroyed by frequent freezing and thawing

Frequent thawing and refreezing of stored antibody samples can lower the titre of some antibody solutions by precipitating immunoglobulin and may encourage dissociation of a label from a conjugated antibody.

(a) Avoid this procedure. Store in small aliquots suitable for one use. Do not freeze slowly by simply placing the sample in a freezer, but 'snap' freeze in liquid nitrogen before storing at −20 °C or below. Rapid thawing is recommended. Store conjugated antibodies at 4 °C.

(b) Do not freeze antibodies in dilute solution. A dilution of up to 1/100 in a buffer solution containing some protective inert protein such as bovine serum albumin at 1% (w/v) is suitable. An alternative is to dilute the antibody 1:1 with glycerol which will allow it to be stored at −20 °C without freezing. The volume of added glycerol must be taken into account in subsequent dilutions. Glycerol will not affect immunoreactivity, but may be removed by dialysis if necessary. Azide may be added but is not necessary for storage at −20 °C.

ii. Dilution

Antibody kept in dilute solution may bind to the walls of the storage vessel, thereby reducing the amount of available antibody. Include a high concentration of an 'inert' protein such as 0.1% bovine serum albumin in the antibody diluent. This competes with the antibody for these non-specific binding sites.

iii. Contamination

(a) Add 0.01−0.1% sodium azide to the antibody diluent to prevent bacterial and fungal growth (care with peroxidase-linked antibodies).

(b) Store antibody in aliquots to avoid frequent opening of bottles. Many antibodies will remain active for several years stored at working dilution in buffer with bovine serum albumin and sodium azide, but this is unpredictable. The immunoreactivity of some antibodies, especially monoclonal ones, may decline rapidly in dilute solution. It is usual to keep immunoreagents at 4 °C but in the author's experience, provided anti-contaminants are included, periods of up to a week at higher temperatures, for transport between laboratories, appear to do no harm.

(c) Store antibody in snap-frozen aliquots, once you have established that it withstands freezing (some monoclonal antibodies are destroyed by freezing).

(d) Immunoglobulins and conjugates may be microfiltered (0.22 µm pore size) and stored in sterile conditions.

iv. Destruction by proteases

Antibodies may be destroyed by active proteases in the antiserum or in any normal serum added to the diluent to provide extra protein. Protein antigens in the tissue or added to the serum in an absorption test (see Section 4.1.2.i) may be similarly affected (52).

(a) Include a protease inhibitor such as Trasylol (Bayer, UK) 4000 U/ml or Bacitracin (Sigma, UK, USA) 100 µg/ml in the antibody diluent (52).

(b) Heat-treat all sera to destroy proteases (this has probably been done by commercial suppliers).

v. Detachment of label from antibody

It is possible that a conjugated label can detach from the antibody during storage. Immunoglobulins released from their label and in solution will bind as readily or better than the labelled molecules and therefore compete for antigenic sites with labelled antibody, reducing the amount of visible reaction. This can be a particular problem for the non-convalently bound colloidal gold-adsorbed antibodies (53).

(a) Do not store labelled antibodies in large quantities. Snap-freeze them (not colloidal gold-linked reagents unless glycerol is added) in small aliquots as soon as you obtain them or follow the supplier's instructions.

(b) Immunogold reagents should be kept at 4 °C and used within their shelf-life (a few months). If glycerol (up to 20%) is added, they may be stored at −20 °C.

(c) Unstabilized gold particles and free or loosely bound antibody can be removed from an immunogold reagent by washing. Centrifuge three times for 1 h at 4 °C in isosmotic-buffered salt solution (20 mM Tris-buffered saline, pH 8.2, containing 1% BSA). The speed of centrifugation depends on the size of the gold particles (60 000g for 5 nm diameter particles, 15 000g for 15−20 nm particles). The gold adsorbed reagent is sedimented as a mobile pool. Remove all but about 10% of the supernatant. Use the remaining supernatant and the loose part of the sediment. The tightly packed sediment consists of aggregated gold and gold conjugate which should be discarded. Further purification to produce a uniform size distribution of gold can be performed by layering the reagent on a sucrose or glycerol density gradient. For details of preparation and purification of immunogold reagents (see 3, 4, 38, and Chapter 8).

vi. *No antibody in the solution*
The possibility exists that the antibody you have elected or been asked to test is not, in fact, a reactive antibody. It may be the product of a non-antibody-secreting clone from an attempt to raise monoclonal antibodies, or a serum from a non-productive immunized animal, or a one-time antibody that has been mistreated and lost its immunoreactivity. In this case, despite the earlier recommendation (3.2.1.iii), test the antibody by an alternative *in vitro* method such as ELISA (54) or dot blotting (55) to check for immunoreactivity before re-trying the possibilities of immunocytochemistry.

3.2.4 Miscellaneous

i. *Fading of fluorescence*
Preparations labelled with a fluorescent marker can fade on exposure to ultraviolet light while being examined or during storage.

(a) Include a fade retarder in the mountant. One such is 1,4-diazobicyclo-(2,2,2)-octane (DABCO) (56) used as 25 mg/ml DABCO (Sigma, UK, USA) in PBS:glycerol 1:9 (v/v). Warm the DABCO in glycerol in a water bath then add to PBS and bring pH to 8.6 with 0.1 M sodium hydroxide. Fluorescence protective mountants are available commercially (Fluormount, Lipshaw, USA, UK).

(b) Repeat the application of the fluorescent conjugate (see 3.1.3(e)).

(c) A fluorescent conjugate can be stabilized against fading by application of an additional layer of antibody raised against the fluorescent label (57). The mechanism is not known.

ii. *No antigen in the tissue*

The tissue you are confidently trying to immunolabel may really contain no antigen. If altering various parameters of the method such as fixation and antibody dilution has no effect:

(a) Try several different antibodies to the same antigen, making sure that they all label positive controls, prepared in the same way as your test tissue.

(b) Try to extract the antigen from your test tissue and stain it in a Western blot, or assay it by RIA, or demonstrate its presence with a chromatographic method.

If these methods fail to demonstrate the antigen, it probably is not in sufficient quantity in the specimen to be detected.

4. Specificity

Confidence in the specific nature of an immunolabelling result is essential. Luckily, diagnostic immunocytochemistry has progressed to the point where the standard localization of commonly detected antigens is well known and antibodies that do not live up to expectation are easily recognized and discarded. In experimental work, however, great care must be taken to demonstrate the specificity of labelling of any newly studied antigen or newly produced antibody.

4.1 Establishment of specificity

It may be worth testing the antibody against its antigen by an alternative method such as ELISA (54) or dot blotting (55) to establish that immuno-reactivity is present. This will not necessarily mean that the antibody is useful immunocytochemically (Section 3.2.1.iii above).

The next stage is to achieve what looks like specific labelling on the tissue preparations that you intend to use.

4.1.1 Non-specific labelling

Eliminate non-specific causes of labelling as outlined in Section 2 above.

4.1.2 Absorption tests

Check that primary antibody is labelling antigen in tissue specifically by pre-absorption with its antigen followed by staining the tissue in the usual way. Binding of the antibody should be eliminated. These tests are used mostly for polyclonal antisera that may contain a variety of antibodies to different portions of the immunogen, as well as antibodies native to the immunized animal. Monoclonal antibodies, by definition, contain a single species of antibody. Provided that they are not binding non-specifically to the tissue, the characterization of their specificity can be done in a tissue-free assay. Cross reactivity (see Section 4.1.4 below) can still be a problem.

i. *Liquid-phase absorption*

Use liquid-phase absorption in the first instance with the antibody at the highest dilution that gives consistent and adequate staining in all positive compartments of the tissue—usually a higher dilution than that which gives the most intense staining. This is to avoid having to add large quantities of expensive antigen to secure total quenching of the antibody. It also ensures that different antibodies can be diluted to approximately the same concentration on the basis of the ratio between the fixed amount of antigen in a particular tissue and an applied antibody. The molecular weight of different antigens is obviously very variable so that 10 µg of one may differ considerably in molar terms from the same weight of another. It is thus useful to express antigen quantities in molar terms so that comparisons between antigens and antibodies can be made (see *Protocols 5* and *6*). Tenfold, stepwise dilution of the antigen, at constant antibody concentration, should allow the labelling to reappear gradually, thus increasing confidence in its specificity. This procedure is not absolutely essential. Many workers merely add to their antibody solution what they hope is an excess of antigen but which may be a vastly extravagant quantity.

Protocol 5. Preparing peptide antigen for liquid-phase absorption test

1. Dissolve the antigen in distilled water containing 1 mg/ml bovine serum albumin (Cohn Fraction V, 99%, Sigma, UK, USA) to protect the antigen and 10 mg/ml lactose (to provide a visible matrix after freezing).

2. Aliquot in amounts containing 1 nmol of antigen in glass vials with rubber stoppers allowing vacuum sealing. Take care to dispense the solution at the bottom of the vial so that it can be re-dissolved in a small volume of antibody.

3. Put stoppers loosely on the vials and lyophilize the solution by putting the vials in a suitable freeze-dryer,[a] at −80 °C for several hours.

4. Evacuate to seal the stoppers. Bring to room temperature.

5. For long term storage keep at −20 °C.

---

[a] If freeze drying facilities are not available, dissolve the antigen in as small a volume as possible, dispense in 1 nmol quantities in vials and snap-freeze. Store frozen. Remember to take the volume of the antigen solution into account when adding antibody for absorption.

The antigen used for absorption must be pure, preferably synthetic, not necessarily the immunogen used to provoke antibody production which might have been an impure extract and coupled with a protein carrier. Antibodies to the impurities or the carrier would be absorbed, giving a false impression of specificity. An unrelated antigen used in parallel should not prevent labelling.

This procedure will confirm that tissue immunoreaction is due to antibody binding to a molecule resembling, in immunological terms, the antigen added and not to non-specific binding of the primary antibody to other tissue components. It will not confirm the absolute identity of the antigen in the tissue which can only be described as 'antigen-like'. The final identification of the antigen could only be proved by biochemical procedures.

Protocol 6. Liquid-phase absorption

1. Determine by a series of immunocytochemical reactions the highest antibody dilution that gives consistent immunolabelling of all positive structures in the tissue.

2. Using this dilution, in standard antibody diluent, add 100 µl antibody to a 1 nmol aliquot of antigen, prepared as in *Protocol 5*. This will give a concentration of 10 nmol of antigen per ml of diluted antibody which should be enough to absorb immunoreactivity. 100 µl of solution is usually an adequate volume for two preparations.

   Take care when releasing the vacuum seal of the antigen vial that the dried antigen powder does not escape. After adding the antibody solution, replace the stopper and seal the vial with Parafilm to prevent evaporation. Roll the vial so that the antibody reaches all surfaces, including the stopper, to dissolve the entire aliquot of antigen.

3. Repeat the procedure to produce a dilution series. Add 100 µl of antibody diluent to another 1 nmol aliquot to give 10 nmol/ml of antigen (solution 1). Add 10 µl of this solution to 100 µl of the diluted antibody. This will give approximately 1 nmol antigen per ml of diluted antibody.

4. Take another 10 µl of the antigen solution (solution 1) and add it to 90 µl of antibody diluent. This is now at 1 nmol of antigen per ml (solution 2). Add 10 µl of this solution to 100 µl of diluted antibody to give approximately 0.1 nmol of antigen per ml of diluted antibody.

5. Take 10 µl of solution 2 and add it to 90 µl of antibody diluent (solution 3, 0.1 nmol of antigen per ml). Add 10 µl of solution 3 to 100 µl of diluted antibody to give 0.01 nmol antigen per ml of diluted antibody.

6. Continue the process as far as 0.001 nmol of antigen per ml of diluted antibody; this is usually sufficient for the reappearance of the immuno-labelling.

Protocol 6. *Continued*

7. Leave the antigen/antibody mixtures to react for 30 min at 37 °C or overnight at 4 °C.

8. Perform the immunolabelling method as usual, using the series of absorbed antibody solutions.

Use a control section from the test material immunolabelled simultaneously with unabsorbed primary antibody at the same dilution as the absorbed antibody.

In addition, if you are checking a new antigen location rather than a new antibody, use the absorbed antibodies in parallel on a known positive control to check that the absorption is satisfactory and that only the specific antigen removes the immunoreactivity from the antibody. Use a positive control with unabsorbed antibody here too.

If immunolabelling is specific and immunoreactivity has been absorbed by adding excess of the antigen to the diluted antibody in solution, no labelling will be seen at the highest antigen concentrations, but it will reappear gradually with diminishing antigen concentrations until it matches the control positive labelling. The same procedure performed with the highest antigen concentration only of any other antigen, whether or not related structurally to the antigen under investigation, should have no effect. The standardization of antigen/antibody mixtures in molar terms allows comparisons to be made between antibodies. The lower the amount of antigen required to 'absorb' an antibody at the limit of its possible dilution, the more specific is that antibody to that antigen.

Liquid-phase absorption may sometimes fail, that is, show no apparent quenching of the antibody reaction with the tissue. Even with high concentrations of antigen, labelling may not be eliminated because of binding of the antigen, used for absorption, to antigen receptors or other sites in the tissue. Since the antigen has previously been attached to the primary antibody, the immunostaining sequence will detect the antibody at the receptor binding site which may well be closely associated with the production or storage site of the antigen. Antigen–antibody complexes formed in solution may also bind to tissue Fc receptors or other attractive sites. These problems are overcome by removing the antibody/antigen complex from the solution using solid-phase absorption.

ii. Solid-phase adsorption

Solid-phase adsorption, combining the antibody with the antigen coupled to a solid support such as cyanogen bromide-activated Sepharose beads, will remove the antibody from solution (*Protocols 7* and *8*) and prevent the hazards of liquid-phase absorption. The beads will bind covalently to proteins or peptides containing amino groups. Disadvantages of this method are that large amounts of antigen are required to coat the beads and it is difficult to know how much is needed or to compare adsorption data from different antibodies and antigens.

Protocol 7. Coupling the antigen to the solid phase (58)

- CNBr-activated Sepharose 4B beads (Pharmacia, UK, Sweden)
- 1 mM HCl
- coupling buffer: 0.1 M NaHCO$_3$, pH 8.3, containing 0.5 M NaCl
- blocking agent: 1 M ethanolamine or 0.2 M glycine, pH 8.0
- acetate buffer, 0.1 M, pH 4.0 containing 0.5 M NaCl

1. Weigh CNBr-activated Sepharose beads. 1 g gives about 3.5 ml final gel.
2. Place the beads on a sintered glass filter (G3 porosity) and add 1 mM HCl to swell the gel. Use 200 ml acid per gram of gel, added in several aliquots. Remove the supernatant between aliquots. Allow the gel to swell for 15 min then wash with 1 mM HCl. The acid prevents hydrolysis of the active groups.
3. Dissolve the antigen in the coupling buffer, 5–10 mg/ml of gel.
4. Wash the gel with coupling buffer and immediately (to avoid hydrolysis of the active groups) add two parts of the antigen solution to one part of gel. Place the mixture on a rotating or other gentle mixer (not a magnetic stirrer, which can cause fragmentation of the beads) for 2 h at room temperature or overnight at 4 °C.
5. Allow to stand, remove the supernatant and add blocking buffer to block remaining active groups. Allow to stand for 2 h at room temperature or overnight at 4 °C.
6. Remove the unadsorbed antigen by several alternate rinses in coupling buffer and acetate buffer.
7. Finally, wash in coupling buffer, add preservative (0.01% sodium azide) and store in aliquots at 4 °C. The gel is now ready to be used for adsorption of the antibody.

Protocol 8. Solid-phase adsorption (58)

The quantity of antigen-bound beads to use per ml of diluted antibody must be determined by experiment. In general, use beads corresponding to up to 100 µg antigen per ml of diluted antibody.

1. Add the coupled beads to the antibody at its limiting dilution (see *Protocol 6*, step **1**).

Protocol 8. *Continued*

2. Mix overnight at 4 °C on a rotary mixer.

3. Allow to stand or centrifuge gently; draw off the supernatant, and use it to stain your preparations. Use control stains as in *Protocol 6*, step **8**.

In theory it would be possible to elute the bound antibody from the solid support with acid buffer and use it as affinity-purified antibody, but in practice only low affinity antibodies are easily eluted and these are not suitable for immunocytochemistry which requires highly avid antibodies.

Solid-phase adsorption is an excellent method for removing known, unwanted, contaminating antibodies from an antiserum. For example, antibodies to the common alpha-subunit of the three pituitary glycoprotein hormones (FSH, LH, TSH) can be adsorbed out of a polyclonal antiserum raised to TSH with alpha-subunit-coated Sepaharose beads to leave only antibodies to the hormone-specific, beta-subunit of TSH.

A simpler, potentially useful method of solid-phase adsorption using antigen bound to the wells of an ELISA plate has been described (59).

4.1.3 Problems of absorption tests

(See reference 5 for a more detailed discussion.) Absorption tests may result in no staining, and thus be thought to have 'worked' if the antibody binds 'non-specifically' to basic proteins. Any antigen containing basic peptide sequences added to such an antibody will therefore result in lack of binding to the tissue, although the blocking is non-specific.

On the other hand, in polyclonal antisera, sub-populations of antibodies that bind specifically to tissue components other than the antigen being investigated will not be removed by absorption with that antigen, which could be confusing. If identified, they could be removed by affinity absorption.

(a) Before an immunocytochemical reaction can be considered specific it must be shown to withstand absorption with a basic protein such as poly-L-lysine (see Section 2.1.6).

(b) The same result should be achieved with several different antibodies to the same antigen.

4.1.4 Cross reactivity

Unwanted specific binding is the most difficult to check for and overcome. The specific, wanted antibody may cross react genuinely with molecules in the tissue that share antigenic sequences with the antigen under consideration. The presence of such molecules may be known and taken into consideration or, more dangerously, be unsuspected. Families of peptides, intermediate filaments and many other groups of substances may share antigenic sequences.

(a) Check for cross reactivity by absorbing the antibody with known related antigens. If absorption takes place, the nature of the antigen labelled in the tissue cannot be absolutely identified by the antibody.

(b) Do not rely solely on immunocytochemistry for antigen identification. The antigen must be identifiable in extracts from the tissue (unless the concentration is too small to be measurable). Use additional methods such as chromatography or immunostained blots of tissue extracts after electrophoresis in a gel to examine the number and molecular size of the immunoreactive antigens in the preparation (60).

(c) Use region-specific antibodies raised to non-cross reacting (as far as known) portions of the molecule. This is the best solution to the problem but such antibodies are not always available.

(d) Use model systems to test antibody specificity and cross reactivity. These can be dot blots of related antigens on nitrocellulose paper immunostained by the antibody using any sensitive method of detection (55), or an ELISA method with antigens adsorbed to walls of plastic wells (54). These methods can also be used to compare the sensitivities of antibodies or detection systems, by using graded concentrations of antigens. It may be necessary to fix and process the substrate-bound antigens in order to simulate, as far as possible, the conditions of an immunocytochemical test for antigen in tissue.

(e) *In situ* hybridization for the mRNA of the antigen (61), if probes are available, can support the immunocytochemical findings (Chapter 9).

5. Multiple immunolabelling

In addition to the possible problems outlined in this chapter, labelling of more than one antigen in a single preparation is vulnerable to the dangers of cross reactivity of anti-species immunoglobulins and, for electron microscopy, of unoccupied binding sites for immunoglobulins on Protein A. These have been discussed in Chapter 6.

6. Conclusions

The length of this chapter may give the impression that immunolabelling is fraught with so many hazards that it is not worth attempting. That this idea is false is shown by the huge employment of the technique in all branches of biological science. Although it is important to be aware of what can go wrong and the possible causes of the disaster, in daily practice few difficulties will be found and, with a systematic approach, most of those that do occur can be overcome.

References

1. Pool, C. W., Buijs, R. M., Swaab, D. F., Boer, G. J., and van Leeuwen, F. W. (1983). In *Immunohistochemistry*, (ed. A. C. Cuello), pp. 1–46. Wiley, Chichester.
2. Sternberger, L. A. (1986). *Immunocytochemistry*, 3rd edn. Wiley, New York.
3. Polak, J. M. and Van Noorden, S. (ed.) (1986). *Immunocytochemistry: Modern Methods and Applications*, 2nd edn. Wright, Bristol.
4. Polak, J. M. and Varndell, I. M. (ed.) (1984). *Immunolabelling for Electron Microscopy*. Elsevier, Amsterdam.
5. Larsson, L.-I. (1988). *Immunocytochemistry: Theory and Practice*. CRC Press, Boca Raton, Florida.
6. Goldenthal, K. L., Hedman, K., Chen, J. W., August, J. T., and Willingham, M. C. (1985). *J. Histochem. Cytochem.*, **33**, 813.
7. Grube, D. (1980). *Histochemistry*, **66**, 149.
8. Bussolati, G. and Gugliotta, P. (1983). *J. Histochem. Cytochem.*, **31**, 1419.
9. Buffa, R., Solcia, E., Fiocca, R., Crivelli, O., and Pera, A. (1979). *J. Histochem. Cytochem.*, **27**, 1279.
10. Krenács, T., Uda, H., and Tanaka, S. (1991). *J. Histochem. Cytochem.*, **39**, 1719.
11. Scopsi, L., Wang, B. L., and Larsson, L.-I. (1986). *J. Histochem. Cytochem.*, **34**, 1469.
12. Wood, G. S. and Warnke, R. (1981). *J. Histochem. Cytochem.*, **29**, 1196.
13. Naritoku, W. Y. and Taylor, C. R. (1982). *J. Histochem. Cytochem.*, **30**, 253.
14. Straus, W. (1972). *J. Histochem. Cytochem.*, **20**, 949.
15. Andrew, S. M. and Jasanin, B. (1987). *Histochem. J.*, **19**, 426.
16. Heyderman, E. (1979). *J. Clin. Pathol.*, **32**, 971.
17. Straus, W. (1983). *Histochemistry*, **78**, 289.
18. Minard, B. J. and Cawley, L. P. (1978). *J. Histochem. Cytochem.*, **26**, 685.
19. Ponder, B. A. and Wilkinson, M. M. (1981). *J. Histochem. Cytochem.*, **29**, 981.
20. Craig, A. S. (1974). *Histochemistry*, **42**, 141.
21. Willingham, M. C. (1983). *J. Histochem. Cytochem.*, **31**, 791.
22. McGovern, J. and Crocker, J. (1986). *J. Clin. Pathol.*, **39**, 923.
23. Danscher, G. and Nørgaard, J. O. R. (1985). *J. Histochem. Cytochem.*, **33**, 76.
24. Gallyas, F. (1982). *Acta Histochem.*, **70**, 99.
25. Gallyas, F. and Wolff, J. R. (1986). *J. Histochem. Cytochem.*, **34**, 1667.
26. Cowen, T., Haven, A. J., and Burnstock, G. (1985). *Histochemistry*, **82**, 205.
27. Khalfan, H., Abuknesha, R., Rand-Weaver, M., Price, R. G., and Robinson, D. (1986). *Histochem. J.*, **18**, 497.
28. Huang, W. M., Gibson, S. J., Facer, P., Gu, J., and Polak, J. M. (1983). *Histochemistry*, **77**, 275.
29. Burns, J. (1987). *J. Clin. Pathol.*, **40**, 858.
30. Weinstein, W. M. and Lechago, J. (1977). *J. Immunol. Methods*, **17**, 375.
31. Finley, J. W. C. and Petrusz, P. (1983). In *Techniques in Immunocytochemistry*, Vol. 1, (ed. G. R. Bullock and P. Petrusz), pp. 239–49. Academic Press, New York.
32. Huang, S., Minassian, H., and More, J. D. (1976). *Lab. Invest.*, **35**, 383.
33. Mepham, B. L., Frater, W., and Mitchell, B. S. (1979). *Histochem. J.*, **11**, 345.
34. Curran, R. C. and Gregory, J. (1980). *J. Clin. Pathol.*, **33**, 1047.
35. Bendayan, M. and Zollinger, M. (1983). *J. Histochem. Cytochem.*, **31**, 101.

36. Falck, B. and Owman, C. A. (1965). *Acta Univ. Lund II*, **7**, 1.
37. Pearse, A. G. E. and Polak. J. M. (1975). *Histochem. J.*, **7**, 179.
38. Beesley, J. E. (1989). *Colloidal Gold: A New Perspective for Cytochemical Marking*. Oxford University Press.
39. Costa, M., Buffa, R., Furness, J. B., and Solcia, E. (1980). *Histochemistry*, **65**, 157.
40. Kuhlman, W. D. and Peschke, P. (1982). *Histochemistry*, **75**, 151.
41. Lane, B. P. and Europa, D. L. (1965). *J. Histochem. Cytochem.*, **13**, 579.
42. Holgate, C. S., Jackson, P., Cowen, P. N., and Bird, C. C. (1983). *J. Histochem. Cytochem.*, **31**, 938.
43. Springall, D. R., Hacker, G. W., Grimelius, L., and Polak, J. M. (1984). *Histochemistry*, **81**, 603.
44. Hacker, G. W., Grimelius, L., Danscher, G., Bernatzky, G., Muss, W., Graf, A. H., *et al.* (1989). *J. Histotechnol.*, **11**, 213.
45. Newman, G. R., Jasani, B., and Williams, E. D. (1983). *J. Microsc.*, **132**, RP1.
46. Peacock, C. S., Thompson, I. W., and Van Noorden, S. (1991). *J. Clin. Pathol.*, **44**, 756.
47. Stierhof, Y. D., Humbel, B. M., and Schwarz, H. (1991). *J. Electron Microsc. Tech.*, **17**, 336.
48. Morris, R. E., Ciraolo, G. M., and Salinger, C. B. (1991). *J. Histochem. Cytochem.*, **39**, 1585.
49. Bigbee, J. W., Kosek, J. C., and Eng, L. E. (1977). *J. Histochem. Cytochem.*, **25**, 443.
50. Gu, J., Islam, K., and Polak, J. M. (1983). *Histochem. J.*, **15**, 475.
51. Ghitescu, L., Galis, Z., and Bendayan, M. (1991). *J. Histochem. Cytochem.*, **39**, 1057.
52. Larsson, L.-I. and Stengaard-Pedersen, K. (1981). *J. Histochem. Cytochem.*, **29**, 1088.
53. Kramarcy, N. A. and Sealock, R. (1991). *J. Histochem. Cytochem.*, **39**, 37.
54. Voller, A., Bidwell, D. W., and Bartlett, A. (1976). *Bull. WHO*, **53**, 55.
55. Scopsi, L. and Larsson, L.-I. (1986). *Histochemistry*, **84**, 221.
56. Johnson, G. D., Davidson, R. S., McNamee, K. C., Russell, G., Goodwin, D., and Holborow, E. J. (1982). *J. Immunol. Methods,* **55**, 231.
57. Abuknesha, R. A., Al-Mazeedi, H. M., and Price, R. G. (1992). *Histochem. J.*, **24**, 73.
58. *Affinity Chromatography, Principles and Methods*. Pharmacia Fine Chemicals, Uppsala.
59. Köves, K. and Arimura, A. (1989). *J. Histochem. Cytochem.*, **37**, 903.
60. Johnstone, A. and Thorpe, R. (1987). *Immunochemistry in Practice*, 2nd edn. Blackwell, Oxford.
61. Penschow, J. D., Haralambidis, J., Darling, P. E., Darby, I. A., Wintour, E. M., Tregear, G. W. *et al.* (1989). In *Regulatory Peptides*, (ed. J. M. Polak), pp. 51–69. Birkhäuser, Basel.

A1

Suppliers of specialist items

This list is divided into (a) suppliers of equipment and general consumables and (b) suppliers of antibodies and probes. Reference has not been made to individual items since product ranges of suppliers are constantly updated and revised.

(a) Suppliers of equipment and general consumables

Agar Scientific Ltd, 66a Cambridge Road, Stansted, Essex, CM24 8DA, UK.
    Agent: Ted Pella Inc., 4595 Mountain Lakes Boulevard, Redding, CA 96003, USA.
    Agent: W. Plannet GmbH, Elektronmikroskopie, Marburger Strasse 90, D-6550 Marburg 7, Germany.

Balzers-BAL-TEC GmbH, im Grohenstuck 2, Walluf, Germany.

Balzers High Vacuum Ltd, Northbridge Road, Berkhamsted, Herts HP4 EN, UK.

BDH, Broom Road, Poole, Dorset BH12 4NN, UK.

Dage-MTI, Michigan City, IN, USA.

Diatome Ltd, 2501 Bienne, Switzerland.

Eastman Kodak Company, Rochester, New York, NY 14652–3512, USA

Graticules Ltd, Tonbridge, Kent TN9 1RM, UK.

Ilford Ltd, Mobberley, Cheshire, UK.

Imaging Technology, Woburn, MA, USA.

Leica UK Ltd, Davy Avenue, Knowlhill, Milton Keynes MK5 8LB, UK.

The London Resin Co. Ltd, PO Box 29, Woking, Surrey GU21 1AE, UK.

National Physical Laboratory, Teddington, Middlesex, TW11 0LW, UK.

Nikon Corporation, Fuji Bldg., Marunouchi 3-chome, Chiyoda-ku, Tokyo 100, Japan.

Nikon Europe B. V., Schipolweg 321, PO Box 222, 1170 AE Badhoevedorp, The Netherlands.

Nikon Inc., 623 Stewart Avenue, Garden City, New York 11530, USA.

Pharmacia, Pharmacia LKB Biotechnology, S-751 82 Uppsala, Sweden.

Pharmacia Ltd, Pharmacia LKB Biotechnology Div., Davy Avenue, Milton Keynes MK5 8PH, UK.

Polysciences Inc., 400 Valley Road, Warrington, PA 18976, USA.

Polysciences Inc., 24 Low Farm Place, Moulton Park, Northampton NN3 1HY, UK.

RMC (Europe) Ltd, PO Box 65, Abingdon, Oxfordshire OX13 5RF, UK.

RMC Inc, 4400 S. Santa Rita Avenue, Tucson, Arizona 85714, USA.

Shandon Inc., 171 Industry Drive, Pittsburgh, PA 15275, USA.

Shandon Scientific Ltd, Astmoor, Runcorn, Cheshire WA7 1PR, UK.

TAAB Laboratories Equipment Ltd, 3 Minerva House, Calleva Industrial Park, Aldermaston, Reading RG7 4QW, UK.

Tousimis Research Corporation, Rochville, MD, USA.

(b) Suppliers of antibodies and probes

Amersham Corp., 2636 S. Clearbrook Drive, Arlington Heights, IL 6005, USA.

Amersham International plc, Lincoln Place, Green End, Aylesbury, HP20 2TP.

Bayer Diagnostics, Evans House, Hamilton Close, Basingstoke, Hampshire RG21 2YE, UK.

Biocell Research Laboratories, Cardiff Business Technology Centre, Senghydd Road, Cardiff CF2 4AY, UK.

BioClinical Services Ltd, Unit 9, Willowbrook Laboratory Units, St Mellons, Cardiff CF3 0EF, UK.

Bioquote Ltd, 3 Mount Pleasant Court, Mount Pleasant, West Yorkshire LS29 8TW, UK.

Bio-Rad Laboratories Inc., 19 Blackstone Street, Cambridge, MA 02139, USA.

Becton and Dickinson Ltd, Between Towns Road, Cowley, Oxford OX4 3IY, UK.

Boehringer Mannheim UK, Diagnostics and Biochemicals Ltd, Bell Lane, Lewes, East Sussex BN7 1LG, UK.

Boehringer Mannheim USA, 7941 Castleway Drive, PO Box 50816, Indianapolis 46250, USA.

British Biotechnology, 4–10 The Quadrant, Barton Lane, Abingdon, OX14 3YS, UK.

Calibiochem, PO Box 12087, San Diego, California 92112–4180, USA.
    Agent: Novabiochem, 3 Heathcote Building, Highfields Science Park, Nottingham NG7 2QH, UK.

Cambridge Research Biochemicals Ltd, Gadbrook Park, Northwich, Cheshire CW9 7RA, UK.

Cambridge Research Biochemicals Ltd, Wilmington, DE 19897, USA.

Cooper Biomedical, Malvern, PA, USA.

DAKO A/S, Produktionsvej 42, PO Box 1359, DK-2600 Glostrup, Denmark.

DAKO Corporation, 22F North Milpas Street, Santa Barbara, California 93103, USA.

DAKO Ltd, 16 Manor Courtyard, Hughenden Avenue, High Wycombe, Bucks HP13 5RE, UK.

E-Y Laboratories Inc., 127 North Amphlett Blvd, San Mateo, CA 94401, USA.

ICN FLOW, ICN Biomedicals Ltd, Eagle House, Peregrine Business Park, Gomm Road, High Wycombe, Bucks HP13 7DL, UK.

ICN Immunobiologicals, Lisle, IL, USA.

Lipshaw, 7446 Central Avenue, Detroit, Michigan 48210, USA.
   Agent: Biogenesis Ltd, 12 Yeomans Park, Yeomans Way, Bournemouth BH8 0BJ, UK.

Miles Scientific, Miles Laboratories Ltd, Stoke Court, Stoke Podges, Slough SL2 4LY, UK.

Nordic Immunological Laboratories, PO Box 544, Maidenhead, Berks SL6 2PW, UK.

Peninsula Laboratories Inc., 611 Taylor Way, Belmont, California 94002, USA.

Peninsula Laboratories Inc., Box 62, 17K Westside Industrial Estate, Jackson Street, St Helens WA9 3AJ, UK.

Pierce, 3747 North Meridian Road, PO Box 117, Rockford, IL 61105, USA.

Sera-Lab Ltd, Crawley Down, West Sussex RH10 4FF, UK.

Serotec Ltd, 22 Bankside Approach, Kidlington, Oxford OX5 1JE, UK.

Sigma Chemical Co., PO Box 14508, St Louis, Missouri 63178, USA.

Sigma Chemical Co. Ltd, Fancy Road, Poole, Dorset BH17 7NH, UK.

Sternberger Meyer Immunocytochemicals Inc., 3739 Jarrettsville Pike, Jarrettsville, MD 21084, USA.

Vector Laboratories, 30 Ingold Road, CA 94010, USA.

Vector Laboratories, 16 Wulfric Square, Bretton, Peterborough PE3 8RF, UK.

Index

absorption tests 231–5
acetone 25
acridine orange 183
adsorption tests 235–6
agarose 31
alcohols 10
aldehydes 8, 9, 10, 22–3, 40, 47
  plant material and 88–9, 91–3, 97
  problems with 215, 222–3
alkaline phosphatase 3
  endogenous 215
  in situ hybridization and 198–9
  multiple immunolabelling and 106
alkaline phosphatase–anti-alkaline
    phosphatase (APAAP) technique 17,
    36–7
ambient-temperature embedding 45–6, 82
amino acids, tissue 212
3-amino-9-ethylcarbazole (AEC) 21, 39, 106
aminopropyltriethoxysilane (APES)-coated
    slides 31, 40–1
anhydrous fixation 88–9, 91–3, 97
antibody (antibodies) 2, 7–9
  carrier protein 212
  cross reactivity of 212, 236–7
  dilution of 288, 227, 229
  hydrophobic bonding and 210–11
  image analysis and 134, 135
  monoclonal 9, 10, 77–8, 79–80, 105, 208,
    222
  multiple immunolabelling and 104–5
    antibody–gold technique 122–3
    labelled antibody technique 114–15
    multiple-antibody technique 123
  plant cells and 77–80
  polyclonal 8–9, 10, 77, 78, 104–5, 208, 222
  problems with 10, 208, 210–11, 212, 217,
    222, 227–30
  validation of 8, 12–13, 44–5, 85, 224, 231–7
antibody–gold technique 122–3
antigen(s) 2, 7
  damage to 10, 43, 44
  epitopes 8, 9
  fixation and 8, 9, 10, 47
  multiple immunolabelling and 103, 104
  phagocytosis of 217
  plant cell 77–80

proteolytic digestion of 23–4, 222–3
  testing of 12–13, 231–7
arabinogalactan 79–80
avidin 18–19
  immunolabelling, saving of 220
  multiple immunolabelling 109–10, 111–12

background labelling 208–18
beta-galactosidase 3, 106
B5 (fixative) 22, 39–40
Bigbee (prozone) effect 227
biotin 18–19, 32–4
  in situ hybridization 187, 189, 194
  multiple immunolabelling 109–12
  problems with 213
blocking agents 12
buffers 12, 226
  hybridization 190–1

cameras 131–2, 136
capping 10
carbohydrate antigens 79
carrier protein antibody 212
cDNA clones 78–9, 177
cell block preparation 31
charge, positive/negative 210–11, 217
chemical reagents 12
4-chloro-1-naphthol 21, 39, 106
chromogens 21, 38–9; *see also individual*
    chromogens
colloidal gold probes 3, 4, 11, 20–1, 70
  aggregate deposition by 217
  correlative microscopic approach 151–2,
    159–74
  image analysis and 148
  in situ hybridization 198
  multiple immunolabelling 106, 108, 110–11,
    117–23
  preparation of 152–9
  protein–gold conjugates 155–9
  silver enhancement of 20–1, 106, 108,
    159–61
    problems with 215–16
    protocols for 37–8, 84–5, 110–11
complement fixation 211–12

controls 8, 12–13, 207–8, 218
  electron microscopy 58, 85–6
  in situ hybridization 200–1
  multiple immunolabelling 105
cross reactivity 212, 236–7
  in multiple immunolabelling 104, 113
cryosection techniques 60–8, 81
cytological preparation techniques 29–31
cytospin preparation 30

dabs 31
dehydration
  electron microscopy 48–9
  light microscopy 24
denaturation technique 121
densitometry studies 135, 137–8, 142–3
  protocol for 145–7
diaminobenzidene (DAB) 21, 40, 106
  problems with 217–18
diamond knives 65
diethylaminocoumarin 106
diethylpyrocarbonate 10, 181–2
digoxigenin 187, 191–2, 194, 195–6, 198–9
2,2-dimethoxypropane (DMP) 88–9, 91–3
N',N'-dimethylformamide 89
direct technique 12, 16
  protocol for 34–5
double-sided technique 121–2

electron microscopy 43–75
  correlative microscopic approach 151–2,
    162–74
  fixation for 8, 9, 47, 49, 62
    plant cell 81–2, 86, 88–9, 91–3, 97
  history of 1, 3–4
  image analysis and 136, 148
  immunolabelling techniques for 43–75
    multiple immunolabelling 118–23
    plant cells 80–97
    post-embedding 43–68, 81–6
    pre-embedding 43, 69–75, 80–1
  in situ hybridization for 196–7
  multiple immunolabelling 103, 104,
    116–23
  probes in 3–4, 11, 70, 116–18
  specimen preparation for 10, 163–4
embedding, *see* resins
enzyme markers 3–4, 11–12, 15
  in situ hybridization and 198–9
  multiple immunolabelling and 105–6
  see also individual enzymes

Fab fragment 8
  capping and 10
Fast Blue BB 106
Fast Red TR 106

Fast Red Violet TB 106
Fc receptors 211
fixation 8, 9, 10
  for electron microscopy 8, 9, 47, 49, 62
  for *in situ* hybridization 182
  for light microscopy 9, 10, 22–3, 25, 39–40
  plant material 81–2, 86, 88–9, 91–3, 97
  problems with 215, 221–4
fluoroscein isothiocyanate 106
fluorescent markers 3, 11, 15
  image analysis 147–8
  in situ hybridization and 198
  problems with 216, 230–1
formaldehyde 9, 10
  vapour 112–13, 121
formalin 10, 22
4-chloro-1-naphthol 21, 39, 106
freeze-substitution 51–60
  of plant cells 93–5
frozen sections 24–5, 33–4, 108
fusion proteins 78–9

glass knives 64–5
glucose oxidase 3, 106
glutaraldehyde 8, 10, 47
glycol methacrylate resins 27
glycoprotein antigens 79–80
gold probes, *see* colloidal gold probes

high voltage transmission electron microscopy
    151–2, 162–74
histology, *see* light microscopy
hybridoma cells 9
hydrophobic bonding 210–11

image analysers 127–31
image analysis 127–49
  automatic 128, 130–1, 138–43
    protocols for 144–7
  equipment for 127–32, 136
    optimization of 137–8
  planning of 132–4
  semi-automatic 128, 143–4
  starting image for 134–6
  techniques for 143–7
immuno-ferritin technique 3, 117
immunofluorescence, *see* fluorescent markers
immunolabelling techniques 11–12
  for electron microscopy 43–75
    multiple immunolabelling 118–23
    plant cells 80–97
    post-embedding 43–68, 81–6
    pre-embedding 43, 69–75, 80–1
  for light microscopy 15–21, 31–8
    multiple immunolabelling 108–15
immunonegative stain technique 72–5

Index

immunoperoxidase, *see* peroxidase
imprints 31
indirect technique 11, 16–17, 35
  multiple immunolabelling 109, 111–12, 114
in situ hybridization 117–203
  problems with 201–2
  technique of 190–7
  theory of 179–80

knives 64–5

labelled antibody technique 114–15
light microscopy 15–41
  correlative microscopic approach 151–2,
    161–2
  fixation for 9, 10, 22–3, 25, 39–41
  history of 1, 3
  image analysis and 132, 136
  immunolabelling techniques for 15–21,
    31–8, 108–15
  multiple immunolabelling 103, 104, 105–15
  probes in 2, 11, 105–8
  of resin sections 26–8, 56–8
  specimen preparation for 10, 23–31
liquid-phase absorption 232–4
lissamine sulphonyl chloride 106
Lowicryl resins 46–50, 55–60
low voltage high resolution scanning electron
  microscopy 151–2, 162–74
LR Gold resin 82
LR White resin 45–6, 82
LX-122 resin 90–1

markers, *see* probe(s)
methanol 10
methylmethacrylate resins 27–8
microscopes, and image analysis 132, 136
monoclonal antibodies 9
  capping and 10
  multiple immunolabelling and 105
  plant cells and 77–8, 79–80
  problems with 208, 222
multiple-antibody technique 123
multiple immunolabelling techniques 103–23
  cross reactions and 104, 113
  electron microscope 103, 104, 116–23
  light microscope 103, 104, 105–15
  of plant cells 98–9
  probes for 103, 105–8, 116–18

necrosis, tissue 217
New Fuchsin 106
nitroblue tetrazolium 106
N',N'-dimethylformamide 89

noise-averaging algorithm 138
nucleic acid probes 177
  hybridization theory 179–80
  isotopic 186–7, 192–4, 200
  labelling of 186–9
  localization of 198–200
  non-isotopic 187–8, 191–2, 194–6, 198–9
  types of 185–6

one step technique 12, 16, 34–5

paraffin sections 23–4, 32–3, 108
paraformaldehyde 22–3, 40, 47
p-benzoquinone 10
periodate lysine paraformaldehyde (PLP) 22,
  40
  dichromate (PLP/D) 22–3, 40
peroxidase 3–4, 16–17, 34–5
  alternative chromagens for 21, 38–9
  endogenous 213–15, 225–6
  multiple immunolabelling with 106, 109
  problems with 213–15, 217, 219–21, 225–6
peroxidase–anti-peroxidase (PAP) technique
  3–4, 11–12, 17
  image analysis and 135
  protocol for 36
  saving of 219–20
phase partition fixation 97
photometric non-uniformity 138
phycoerythrin 106
pigment in tissue 215
plant cells 77–100
  antigens in 77–80
  immunolabelling techniques for 80–6
  multiple immunolabelling in 98–9
  protocols for 86–97
  transgenic 100
polyclonal sera 8–9
  capping and 10
  multiple immunolabelling and 104–5
  plant cells and 77, 78
  problems with 208, 222
polymerase chain reaction 203
potassium ferro/ferricyanide 106
probe(s) 2, 11
  colloidal gold 3, 4, 11, 20–1, 70
    correlative microscopic approach 151–2,
     159–74
    image analysis and 148
    in situ hybridization 198
    multiple immunolabelling 106, 108,
     110–11, 117–23
    preparation of 152–9
    problems with 215–16
  enzyme 3–4, 11–12, 15, 105–6
    in situ hybridization and 198–9
    see also individual enzymes

probe(s) (*cont..*):
  fluorescent 3, 11, 15, 105–6
    image analysis and 147–8
    in situ hybridization and 198
    problems with 216, 230–1
  multiple immunolabelling 103, 105–8,
      110–11, 116–23
  nucleic acid 177, 179–80
    isotopic 186–7, 192–4, 200
    labelling of 186–9
    localization of 198–200
    non-isotopic 187–8, 191–2, 194–6, 198–9
    types of 185–6
  problems 207–37
    background labelling 208–18
    in situ hybridization 201–2
    method problems 218–31
    specificity tests 231–7
progressive lowering of temperature (PLT)
      embedding 46–50, 55–60, 83
Protein A 4, 11, 17, 35–6
  multiple immunolabelling 119–20
  problems with 228
Protein G 4, 11
protein–gold conjugates 155–9

recombinant proteins 78–9
resins
  for electron microscopy 45–60, 121–2
    plant cells 80–97
    hydrophobic 90–1
  for light microscopy 26–8, 56–8, 108
  for multiple immunolabelling 108, 121–2
RNase-free environment 180–2

shade correction algorithms 138

silver enhancement 20–1, 106, 108, 159–61
  problems with 215–16
  protocols for 37–8, 84–5, 110–11, 120
smear preparation 30
solid-phase adsorption 234
specificity tests 231–7
specimen preparation 9–10
  for correlative microscopic approach 163–4
  for electron microscopy 10, 163–4
  for *in situ* hybridization 182–5
  for light microscopy 10, 23–31
  problems 221–6
steric hindrance 105, 117, 228
storage 228–30
streptavidin 4, 11, 19, 32–4
  multiple immunlabelling 110–11
supraoptimal dilution technique 135
surfactant reagents 12

temporal drift 137
TESPA-coated slides 182–3
tetramethylrhodamine-isothiocyanate 106
Texas red 106
thawed crysection techniques 60–8
3-amino-9-ethylcarbazole (AEC) 21, 39, 106
three step technique 11–12
  multiple immunolabelling 109–10
transgenic plants 100
trypsin 23–4, 222–3
two step (indirect) technique 11, 16–17, 35
  multiple immunolabelling 109, 111–12, 114
2,2-dimethoxypropane (DMP) 88–9, 91–3

vapour fixation 91, 97, 112–13, 121
video-enhanced differential interference
      contrast light microscopy 151–2, 161–2

ORDER OTHER TITLES OF INTEREST TODAY

Price list for: UK, Europe, Rest of World (excluding US and Canada)

Forthcoming Titles

124. Human Genetic Disease Analysis Davies, K.E. (Ed)
...... Spiralbound hardback 0-19-963309-6 **£30.00**
...... Paperback 0-19-963308-8 **£18.50**
123. Protein Phosphorylation Hardie, G. (Ed)
...... Spiralbound hardback 0-19-963306-1 **£32.50**
...... Paperback 0-19-963305-3 **£22.50**
122. Immunocytochemistry Beesley, J. (Ed)
...... Spiralbound hardback 0-19-963270-7 **£32.50**
...... Paperback 0-19-963269-3 **£22.50**
121. Tumour Immunobiology Gallagher, G., Rees, R.C. & others (Eds)
...... Spiralbound hardback 0-19-963370-3 **£35.00**
...... Paperback 0-19-963369-X **£25.00**
120. Transcription Factors Latchman, D.S. (Ed)
...... Spiralbound hardback 0-19-963342-8 **£30.00**
...... Paperback 0-19-963341-X **£19.50**
119. Growth Factors McKay, I.A. & Leigh, I. (Eds)
...... Spiralbound hardback 0-19-963360-6 **£30.00**
...... Paperback 0-19-963359-2 **£19.50**
118. Histocompatibility Testing Dyer, P. & Middleton, D. (Eds)
...... Spiralbound hardback 0-19-963364-9 **£32.50**
...... Paperback 0-19-963363-0 **£22.50**
117. Gene Transcription Hames, D.B. & Higgins, S.J. (Eds)
...... Spiralbound hardback 0-19-963292-8 **£35.00**
...... Paperback 0-19-963291-X **£25.00**
116. Electrophysiology Wallis, D.I. (Ed)
...... Spiralbound hardback 0-19-963348-7 **£32.50**
...... Paperback 0-19-963347-9 **£22.50**
115. Biological Data Analysis Fry, J.C. (Ed)
...... Spiralbound hardback 0-19-963340-1 **£50.00**
...... Paperback 0-19-963339-8 **£27.50**
114. Experimental Neuroanatomy Bolam, J.P. (Ed)
...... Spiralbound hardback 0-19-963326-6 **£32.50**
...... Paperback 0-19-963325-8 **£22.50**
112. Lipid Analysis Hamilton, R.J. & Hamilton, S.J. (Eds)
...... Spiralbound hardback 0-19-963098-4 **£35.00**
...... Paperback 0-19-963099-2 **£25.00**
111. Haemopoiesis Testa, N.G. & Molineux, G. (Eds)
...... Spiralbound hardback 0-19-963366-5 **£32.50**
...... Paperback 0-19-963365-7 **£22.50**

Published Titles

113. Preparative Centrifugation Rickwood, D. (Ed)
...... Spiralbound hardback 0-19-963208-1 **£45.00**
...... Paperback 0-19-963211-1 **£25.00**
110. Pollination Ecology Dafni, A.
...... Spiralbound hardback 0-19-963299-5 **£32.50**
...... Paperback 0-19-963298-7 **£22.50**
109. In Situ Hybridization Wilkinson, D.G. (Ed)
...... Spiralbound hardback 0-19-963328-2 **£30.00**
...... Paperback 0-19-963327-4 **£18.50**
108. Protein Engineering Rees, A.R., Sternberg, M.J.E. & others (Eds)
...... Spiralbound hardback 0-19-963139-5 **£35.00**
...... Paperback 0-19-963138-7 **£25.00**

107. Cell-Cell Interactions Stevenson, B.R., Gallin, W.J. & others (Eds)
...... Spiralbound hardback 0-19-963319-3 **£32.50**
...... Paperback 0-19-963318-5 **£22.50**
106. Diagnostic Molecular Pathology: Volume I Herrington, C.S. & McGee, J. O'D. (Eds)
...... Spiralbound hardback 0-19-963237-5 **£30.00**
...... Paperback 0-19-963236-7 **£19.50**
105. Biomechanics-Materials Vincent, J.F.V. (Ed)
...... Spiralbound hardback 0-19-963223-5 **£35.00**
...... Paperback 0-19-963222-7 **£25.00**
104. Animal Cell Culture (2/e) Freshney, R.I. (Ed)
...... Spiralbound hardback 0-19-963212-X **£30.00**
...... Paperback 0-19-963213-8 **£19.50**
103. Molecular Plant Pathology: Volume II Gurr, S.J., McPherson, M.J. & others (Eds)
...... Spiralbound hardback 0-19-963352-5 **£32.50**
...... Paperback 0-19-963351-7 **£22.50**
101. Protein Targeting Magee, A.I. & Wileman, T. (Eds)
...... Spiralbound hardback 0-19-963206-5 **£32.50**
...... Paperback 0-19-963210-3 **£22.50**
100. Diagnostic Molecular Pathology: Volume II: Cell and Tissue Genotyping Herrington, C.S. & McGee, J.O'D. (Eds)
...... Spiralbound hardback 0-19-963239-1 **£30.00**
...... Paperback 0-19-963238-3 **£19.50**
99. Neuronal Cell Lines Wood, J.N. (Ed)
...... Spiralbound hardback 0-19-963346-0 **£32.50**
...... Paperback 0-19-963345-2 **£22.50**
98. Neural Transplantation Dunnett, S.B. & Björklund, A. (Eds)
...... Spiralbound hardback 0-19-963286-3 **£30.00**
...... Paperback 0-19-963285-5 **£19.50**
97. Human Cytogenetics: Volume II: Malignancy and Acquired Abnormalities (2/e) Rooney, D.E. & Czepulkowski, B.H. (Eds)
...... Spiralbound hardback 0-19-963290-1 **£30.00**
...... Paperback 0-19-963289-8 **£22.50**
96. Human Cytogenetics: Volume I: Constitutional Analysis (2/e) Rooney, D.E. & Czepulkowski, B.H. (Eds)
...... Spiralbound hardback 0-19-963288-X **£30.00**
...... Paperback 0-19-963287-1 **£22.50**
95. Lipid Modification of Proteins Hooper, N.M. & Turner, A.J. (Eds)
...... Spiralbound hardback 0-19-963274-X **£32.50**
...... Paperback 0-19-963273-1 **£22.50**
94. Biomechanics-Structures and Systems Biewener, A.A. (Ed)
...... Spiralbound hardback 0-19-963268-5 **£42.50**
...... Paperback 0-19-963267-7 **£25.00**
93. Lipoprotein Analysis Converse, C.A. & Skinner, E.R. (Eds)
...... Spiralbound hardback 0-19-963192-1 **£30.00**
...... Paperback 0-19-963231-6 **£19.50**
92. Receptor-Ligand Interactions Hulme, E.C. (Ed)
...... Spiralbound hardback 0-19-963090-9 **£35.00**
...... Paperback 0-19-963091-7 **£25.00**
91. Molecular Genetic Analysis of Populations Hoelzel, A.R. (Ed)
...... Spiralbound hardback 0-19-963278-2 **£32.50**
...... Paperback 0-19-963277-4 **£22.50**

90. **Enzyme Assays** Eisenthal, R. & Danson, M.J. (Eds)
...... Spiralbound hardback 0-19-963142-5 **£35.00**
...... Paperback 0-19-963143-3 **£25.00**

89. **Microcomputers in Biochemistry** Bryce, C.F.A. (Ed)
...... Spiralbound hardback 0-19-963253-7 **£30.00**
...... Paperback 0-19-963252-9 **£19.50**

88. **The Cytoskeleton** Carraway, K.L. & Carraway, C.A.C. (Eds)
...... Spiralbound hardback 0-19-963257-X **£30.00**
...... Paperback 0-19-963256-1 **£19.50**

87. **Monitoring Neuronal Activity** Stamford, J.A. (Ed)
...... Spiralbound hardback 0-19-963244-8 **£30.00**
...... Paperback 0-19-963243-X **£19.50**

86. **Crystallization of Nucleic Acids and Proteins** Ducruix, A. & Giegé, R. (Eds)
...... Spiralbound hardback 0-19-963245-6 **£35.00**
...... Paperback 0-19-963246-4 **£25.00**

85. **Molecular Plant Pathology: Volume I** Gurr, S.J., McPherson, M.J. & others (Eds)
...... Spiralbound hardback 0-19-963103-4 **£30.00**
...... Paperback 0-19-963102-6 **£19.50**

84. **Anaerobic Microbiology** Levett, P.N. (Ed)
...... Spiralbound hardback 0-19-963204-9 **£32.50**
...... Paperback 0-19-963262-6 **£22.50**

83. **Oligonucleotides and Analogues** Eckstein, F. (Ed)
...... Spiralbound hardback 0-19-963280-4 **£32.50**
...... Paperback 0-19-963279-0 **£22.50**

82. **Electron Microscopy in Biology** Harris, R. (Ed)
...... Spiralbound hardback 0-19-963219-7 **£32.50**
...... Paperback 0-19-963215-4 **£22.50**

81. **Essential Molecular Biology: Volume II** Brown, T.A. (Ed)
...... Spiralbound hardback 0-19-963112-3 **£32.50**
...... Paperback 0-19-963113-1 **£22.50**

80. **Cellular Calcium** McCormack, J.G. & Cobbold, P.H. (Eds)
...... Spiralbound hardback 0-19-963131-X **£35.00**
...... Paperback 0-19-963130-1 **£25.00**

79. **Protein Architecture** Lesk, A.M.
...... Spiralbound hardback 0-19-963054-2 **£32.50**
...... Paperback 0-19-963055-0 **£22.50**

78. **Cellular Neurobiology** Chad, J. & Wheal, H. (Eds)
...... Spiralbound hardback 0-19-963106-9 **£32.50**
...... Paperback 0-19-963107-7 **£22.50**

77. **PCR** McPherson, M.J., Quirke, P. & others (Eds)
...... Spiralbound hardback 0-19-963226-X **£30.00**
...... Paperback 0-19-963196-4 **£19.50**

76. **Mammalian Cell Biotechnology** Butler, M. (Ed)
...... Spiralbound hardback 0-19-963207-3 **£30.00**
...... Paperback 0-19-963209-X **£19.50**

75. **Cytokines** Balkwill, F.R. (Ed)
...... Spiralbound hardback 0-19-963218-9 **£35.00**
...... Paperback 0-19-963214-6 **£25.00**

74. **Molecular Neurobiology** Chad, J. & Wheal, H. (Eds)
...... Spiralbound hardback 0-19-963108-5 **£30.00**
...... Paperback 0-19-963109-3 **£19.50**

73. **Directed Mutagenesis** McPherson, M.J. (Ed)
...... Spiralbound hardback 0-19-963141-7 **£30.00**
...... Paperback 0-19-963140-9 **£19.50**

72. **Essential Molecular Biology: Volume I** Brown, T.A. (Ed)
...... Spiralbound hardback 0-19-963110-7 **£32.50**
...... Paperback 0-19-963111-5 **£22.50**

71. **Peptide Hormone Action** Siddle, K. & Hutton, J.C.
...... Spiralbound hardback 0-19-963070-4 **£32.50**
...... Paperback 0-19-963071-2 **£22.50**

70. **Peptide Hormone Secretion** Hutton, J.C. & Siddle, K. (Eds)
...... Spiralbound hardback 0-19-963068-2 **£35.00**
...... Paperback 0-19-963069-0 **£25.00**

69. **Postimplantation Mammalian Embryos** Copp, A.J. & Cockroft, D.L. (Eds)
...... Spiralbound hardback 0-19-963088-7 **£35.00**
...... Paperback 0-19-963089-5 **£25.00**

68. **Receptor-Effector Coupling** Hulme, E.C. (Ed)
...... Spiralbound hardback 0-19-963094-1 **£30.00**
...... Paperback 0-19-963095-X **£19.50**

67. **Gel Electrophoresis of Proteins (2/e)** Hames, B.D. & Rickwood, D. (Eds)
...... Spiralbound hardback 0-19-963074-7 **£35.00**
...... Paperback 0-19-963075-5 **£25.00**

66. **Clinical Immunology** Gooi, H.C. & Chapel, H. (Eds)
...... Spiralbound hardback 0-19-963086-0 **£32.50**
...... Paperback 0-19-963087-9 **£22.50**

65. **Receptor Biochemistry** Hulme, E.C. (Ed)
...... Spiralbound hardback 0-19-963092-5 **£35.00**
...... Paperback 0-19-963093-3 **£25.00**

64. **Gel Electrophoresis of Nucleic Acids (2/e)** Rickwood, D. & Hames, B.D. (Eds)
...... Spiralbound hardback 0-19-963082-8 **£32.50**
...... Paperback 0-19-963083-6 **£22.50**

63. **Animal Virus Pathogenesis** Oldstone, M.B.A. (Ed)
...... Spiralbound hardback 0-19-963100-X **£30.00**
...... Paperback 0-19-963101-8 **£18.50**

62. **Flow Cytometry** Ormerod, M.G. (Ed)
...... Paperback 0-19-963053-4 **£22.50**

61. **Radioisotopes in Biology** Slater, R.J. (Ed)
...... Spiralbound hardback 0-19-963080-1 **£32.50**
...... Paperback 0-19-963081-X **£22.50**

60. **Biosensors** Cass, A.E.G. (Ed)
...... Spiralbound hardback 0-19-963046-1 **£30.00**
...... Paperback 0-19-963047-X **£19.50**

59. **Ribosomes and Protein Synthesis** Spedding, G. (Ed)
...... Spiralbound hardback 0-19-963104-2 **£32.50**
...... Paperback 0-19-963105-0 **£22.50**

58. **Liposomes** New, R.R.C. (Ed)
...... Spiralbound hardback 0-19-963076-3 **£35.00**
...... Paperback 0-19-963077-1 **£22.50**

57. **Fermentation** McNeil, B. & Harvey, L.M. (Eds)
...... Spiralbound hardback 0-19-963044-5 **£30.00**
...... Paperback 0-19-963045-3 **£19.50**

56. **Protein Purification Applications** Harris, E.L.V. & Angal, S. (Eds)
...... Spiralbound hardback 0-19-963022-4 **£30.00**
...... Paperback 0-19-963023-2 **£18.50**

55. **Nucleic Acids Sequencing** Howe, C.J. & Ward, E.S. (Eds)
...... Spiralbound hardback 0-19-963056-9 **£30.00**
...... Paperback 0-19-963057-7 **£19.50**

54. **Protein Purification Methods** Harris, E.L.V. & Angal, S. (Eds)
...... Spiralbound hardback 0-19-963002-X **£30.00**
...... Paperback 0-19-963003-8 **£20.00**

53. **Solid Phase Peptide Synthesis** Atherton, E. & Sheppard, R.C.
...... Spiralbound hardback 0-19-963066-6 **£30.00**
...... Paperback 0-19-963067-4 **£18.50**

52. **Medical Bacteriology** Hawkey, P.M. & Lewis, D.A. (Eds)
...... Spiralbound hardback 0-19-963008-9 **£38.00**
...... Paperback 0-19-963009-7 **£25.00**

51. **Proteolytic Enzymes** Beynon, R.J. & Bond, J.S. (Eds)
...... Spiralbound hardback 0-19-963058-5 **£30.00**
...... Paperback 0-19-963059-3 **£19.50**

50. **Medical Mycology** Evans, E.G.V. & Richardson, M.D. (Eds)
...... Spiralbound hardback 0-19-963010-0 **£37.50**
...... Paperback 0-19-963011-9 **£25.00**

49. **Computers in Microbiology** Bryant, T.N. & Wimpenny, J.W.T. (Eds)
...... Paperback 0-19-963015-1 **£19.50**

48. **Protein Sequencing** Findlay, J.B.C. & Geisow, M.J. (Eds)
...... Spiralbound hardback 0-19-963012-7 **£30.00**
...... Paperback 0-19-963013-5 **£18.50**

47. **Cell Growth and Division** Baserga, R. (Ed)
...... Spiralbound hardback 0-19-963026-7 **£30.00**
...... Paperback 0-19-963027-5 **£18.50**

46. **Protein Function** Creighton, T.E. (Ed)
...... Spiralbound hardback 0-19-963006-2 **£32.50**
...... Paperback 0-19-963007-0 **£22.50**

45. **Protein Structure** Creighton, T.E. (Ed)
...... Spiralbound hardback 0-19-963000-3 **£32.50**
...... Paperback 0-19-963001-1 **£22.50**

44. **Antibodies: Volume II** Catty, D. (Ed)
...... Spiralbound hardback 0-19-963018-6 **£30.00**
...... Paperback 0-19-963019-4 **£19.50**

43. **HPLC of Macromolecules** Oliver, R.W.A. (Ed)
...... Spiralbound hardback 0-19-963020-8 **£30.00**
...... Paperback 0-19-963021-6 **£19.50**
42. **Light Microscopy in Biology** Lacey, A.J. (Ed)
...... Spiralbound hardback 0-19-963036-4 **£30.00**
...... Paperback 0-19-963037-2 **£19.50**
41. **Plant Molecular Biology** Shaw, C.H. (Ed)
...... Paperback 1-85221-056-7 **£22.50**
40. **Microcomputers in Physiology** Fraser, P.J. (Ed)
...... Spiralbound hardback 1-85221-129-6 **£30.00**
...... Paperback 1-85221-130-X **£19.50**
39. **Genome Analysis** Davies, K.E. (Ed)
...... Spiralbound hardback 1-85221-109-1 **£30.00**
...... Paperback 1-85221-110-5 **£18.50**
38. **Antibodies: Volume I** Catty, D. (Ed)
...... Paperback 0-947946-85-3 **£19.50**
37. **Yeast** Campbell, I. & Duffus, J.H. (Eds)
...... Paperback 0-947946-79-9 **£19.50**
36. **Mammalian Development** Monk, M. (Ed)
...... Hardback 1-85221-030-3 **£30.00**
...... Paperback 1-85221-029-X **£22.50**
35. **Lymphocytes** Klaus, G.G.B. (Ed)
...... Hardback 1-85221-018-4 **£30.00**
34. **Lymphokines and Interferons** Clemens, M.J., Morris, A.G. & others (Eds)
...... Paperback 1-85221-035-4 **£22.50**
33. **Mitochondria** Darley-Usmar, V.M., Rickwood, D. & others (Eds)
...... Hardback 1-85221-034-6 **£32.50**
...... Paperback 1-85221-033-8 **£22.50**
32. **Prostaglandins and Related Substances** Benedetto, C., McDonald-Gibson, R.G. & others (Eds)
...... Hardback 1-85221-032-X **£32.50**
...... Paperback 1-85221-031-1 **£22.50**
31. **DNA Cloning: Volume III** Glover, D.M. (Ed)
...... Hardback 1-85221-049-4 **£30.00**
...... Paperback 1-85221-048-6 **£19.50**
30. **Steroid Hormones** Green, B. & Leake, R.E. (Eds)
...... Paperback 0-947946-53-5 **£19.50**
29. **Neurochemistry** Turner, A.J. & Bachelard, H.S. (Eds)
...... Hardback 1-85221-028-1 **£30.00**
...... Paperback 1-85221-027-3 **£19.50**
28. **Biological Membranes** Findlay, J.B.C. & Evans, W.H. (Eds)
...... Hardback 0-947946-84-5 **£32.50**
...... Paperback 0-947946-83-7 **£22.50**
27. **Nucleic Acid and Protein Sequence Analysis** Bishop, M.J. & Rawlings, C.J. (Eds)
...... Hardback 1-85221-007-9 **£35.00**
...... Paperback 1-85221-006-0 **£25.00**
26. **Electron Microscopy in Molecular Biology** Sommerville, J. & Scheer, U. (Eds)
...... Hardback 0-947946-64-0 **£30.00**
...... Paperback 0-947946-54-3 **£19.50**
25. **Teratocarcinomas and Embryonic Stem Cells** Robertson, E.J. (Ed)
...... Hardback 1-85221-005-2 **£19.50**
...... Paperback 1-85221-004-4 **£19.50**
24. **Spectrophotometry and Spectrofluorimetry** Harris, D.A. & Bashford, C.L. (Eds)
...... Hardback 0-947946-69-1 **£30.00**
...... Paperback 0-947946-46-2 **£18.50**
23. **Plasmids** Hardy, K.G. (Ed)
...... Paperback 0-947946-81-0 **£18.50**
22. **Biochemical Toxicology** Snell, K. & Mullock, B. (Eds)
...... Paperback 0-947946-52-7 **£19.50**
19. **Drosophila** Roberts, D.B. (Ed)
...... Hardback 0-947946-66-7 **£32.50**
...... Paperback 0-947946-45-4 **£22.50**
17. **Photosynthesis: Energy Transduction** Hipkins, M.F. & Baker, N.R. (Eds)
...... Hardback 0-947946-63-2 **£30.00**
...... Paperback 0-947946-51-9 **£18.50**
16. **Human Genetic Diseases** Davies, K.E. (Ed)
...... Hardback 0-947946-76-4 **£30.00**
...... Paperback 0-947946-75-6 **£18.50**

14. **Nucleic Acid Hybridisation** Hames, B.D. & Higgins, S.J. (Eds)
...... Hardback 0-947946-61-6 **£30.00**
...... Paperback 0-947946-23-3 **£19.50**
13. **Immobilised Cells and Enzymes** Woodward, J. (Ed)
...... Hardback 0-947946-60-8 **£18.50**
12. **Plant Cell Culture** Dixon, R.A. (Ed)
...... Paperback 0-947946-22-5 **£19.50**
11a. **DNA Cloning: Volume I** Glover, D.M. (Ed)
...... Paperback 0-947946-18-7 **£18.50**
11b. **DNA Cloning: Volume II** Glover, D.M. (Ed)
...... Paperback 0-947946-19-5 **£19.50**
10. **Virology** Mahy, B.W.J. (Ed)
...... Paperback 0-904147-78-9 **£19.50**
9. **Affinity Chromatography** Dean, P.D.G., Johnson, W.S. & others (Eds)
...... Paperback 0-904147-71-1 **£19.50**
7. **Microcomputers in Biology** Ireland, C.R. & Long, S.P. (Eds)
...... Paperback 0-904147-57-6 **£18.00**
6. **Oligonucleotide Synthesis** Gait, M.J. (Ed)
...... Paperback 0-904147-74-6 **£18.50**
5. **Transcription and Translation** Hames, B.D. & Higgins, S.J. (Eds)
...... Paperback 0-904147-52-5 **£22.50**
3. **Iodinated Density Gradient Media** Rickwood, D. (Ed)
...... Paperback 0-904147-51-7 **£19.50**

Sets

Essential Molecular Biology: Volumes I and II as a set Brown, T.A. (Ed)
...... Spiralbound hardback 0-19-963114-X **£58.00**
...... Paperback 0-19-963115-8 **£40.00**
Antibodies: Volumes I and II as a set Catty, D. (Ed)
...... Paperback 0-19-963063-1 **£33.00**
Cellular and Molecular Neurobiology Chad, J. & Wheal, H. (Eds)
...... Spiralbound hardback 0-19-963255-3 **£56.00**
...... Paperback 0-19-963254-5 **£38.00**
Protein Structure and Protein Function: Two-volume set Creighton, T.E. (Ed)
...... Spiralbound hardback 0-19-963064-X **£55.00**
...... Paperback 0-19-963065-8 **£38.00**
DNA Cloning: Volumes I, II, III as a set Glover, D.M. (Ed)
...... Paperback 1-85221-069-9 **£46.00**
Molecular Plant Pathology: Volumes I and II as a set Gurr, S.J., McPherson, M.J. & others (Eds)
...... Spiralbound hardback 0-19-963354-1 **£56.00**
...... Paperback 0-19-963353-3 **£37.00**
Protein Purification Methods, and Protein Purification Applications, two-volume set Harris, E.L.V. & Angal, S. (Eds)
...... Spiralbound hardback 0-19-963048-8 **£48.00**
...... Paperback 0-19-963049-6 **£32.00**
Diagnostic Molecular Pathology: Volumes I and II as a set Herrington, C.S. & McGee, J. O'D. (Eds)
...... Spiralbound hardback 0-19-963241-3 **£54.00**
...... Paperback 0-19-963240-5 **£35.00**
Receptor Biochemistry; Receptor-Effector Coupling; Receptor-Ligand Interactions Hulme, E.C. (Ed)
...... Spiralbound hardback 0-19-963096-8 **£90.00**
...... Paperback 0-19-963097-6 **£62.50**
Signal Transduction Milligan, G. (Ed)
...... Spiralbound hardback 0-19-963296-0 **£30.00**
...... Paperback 0-19-963295-2 **£18.50**
Human Cytogenetics: Volumes I and II as a set (2/e) Rooney, D.E. & Czepulkowski, B.H. (Eds)
...... Hardback 0-19-963314-2 **£58.50**
...... Paperback 0-19-963313-4 **£40.50**
Peptide Hormone Secretion/Peptide Hormone Action Siddle, K. & Hutton, J.C. (Eds)
...... Spiralbound hardback 0-19-963072-0 **£55.00**
...... Paperback 0-19-963073-9 **£38.00**

ORDER FORM for UK, Europe and Rest of World

(Excluding USA and Canada)

Qty	ISBN	Author	Title	Amount
			P&P	
			TOTAL	

Please add postage and packing: £1.75 for UK orders under £20; £2.75 for UK orders over £20; overseas orders add 10% of total.

Name ....

Address ....

....

.... Post code ....

[ ] Please charge £ .... to my credit card
Access/VISA/Eurocard/AMEX/Diners Club (circle appropriate card)

Card No .... Expiry date ....

Signature ....

Credit card account address if different from above:

....

.... Postcode ....

[ ] I enclose a cheque for £......................

Please return this form to: OUP Distribution Services, Saxon Way West, Corby, Northants NN18 9ES

OR ORDER BY CREDIT CARD HOTLINE: Tel +44-(0)536-741519 or Fax +44-(0)536-746337

ORDER OTHER TITLES OF
INTEREST TODAY

Price list for: USA and Canada

123.	**Protein Phosphorylation** Hardie, G. (Ed)		
......	Spiralbound hardback	0-19-963306-1	**$65.00**
......	Paperback	0-19-963305-3	**$45.00**
121.	**Tumour Immunobiology** Gallagher, G., Rees, R.C. & others (Eds)		
......	Spiralbound hardback	0-19-963370-3	**$72.00**
......	Paperback	0-19-963369-X	**$50.00**
117.	**Gene Transcription** Hames, D.B. & Higgins, S.J. (Eds)		
......	Spiralbound hardback	0-19-963292-8	**$72.00**
......	Paperback	0-19-963291-X	**$50.00**
116.	**Electrophysiology** Wallis, D.I. (Ed)		
......	Spiralbound hardback	0-19-963348-7	**$66.50**
......	Paperback	0-19-963347-9	**$45.95**
115.	**Biological Data Analysis** Fry, J.C. (Ed)		
......	Spiralbound hardback	0-19-963340-1	**$80.00**
......	Paperback	0-19-963339-8	**$60.00**
114.	**Experimental Neuroanatomy** Bolam, J.P. (Ed)		
......	Spiralbound hardback	0-19-963326-6	**$65.00**
......	Paperback	0-19-963325-8	**$40.00**
111.	**Haemopoiesis** Testa, N.G. & Molineux, G. (Eds)		
......	Spiralbound hardback	0-19-963366-5	**$65.00**
......	Paperback	0-19-963365-7	**$45.00**
113.	**Preparative Centrifugation** Rickwood, D. (Ed)		
......	Spiralbound hardback	0-19-963208-1	**$90.00**
......	Paperback	0-19-963211-1	**$50.00**
110.	**Pollination Ecology** Dafni, A.		
......	Spiralbound hardback	0-19-963299-5	**$65.00**
......	Paperback	0-19-963298-7	**$45.00**
109.	**In Situ Hybridization** Wilkinson, D.G. (Ed)		
......	Spiralbound hardback	0-19-963328-2	**$58.00**
......	Paperback	0-19-963327-4	**$36.00**
108.	**Protein Engineering** Rees, A.R., Sternberg, M.J.E. & others (Eds)		
......	Spiralbound hardback	0-19-963139-5	**$75.00**
......	Paperback	0-19-963138-7	**$50.00**
107.	**Cell-Cell Interactions** Stevenson, B.R., Gallin, W.J. & others (Eds)		
......	Spiralbound hardback	0-19-963319-3	**$60.00**
......	Paperback	0-19-963318-5	**$40.00**
106.	**Diagnostic Molecular Pathology: Volume I** Herrington, C.S. & McGee, J. O'D. (Eds)		
......	Spiralbound hardback	0-19-963237-5	**$58.00**
......	Paperback	0-19-963236-7	**$38.00**
105.	**Biomechanics-Materials** Vincent, J.F.V. (Ed)		
......	Spiralbound hardback	0-19-963223-5	**$70.00**
......	Paperback	0-19-963222-7	**$50.00**
104.	**Animal Cell Culture (2/e)** Freshney, R.I. (Ed)		
......	Spiralbound hardback	0-19-963212-X	**$60.00**
......	Paperback	0-19-963213-8	**$40.00**
103.	**Molecular Plant Pathology: Volume II** Gurr, S.J., McPherson, M.J. & others (Eds)		
......	Spiralbound hardback	0-19-963352-5	**$65.00**
......	Paperback	0-19-963351-7	**$45.00**
101.	**Protein Targeting** Magee, A.I. & Wileman, T. (Eds)		
......	Spiralbound hardback	0-19-963206-5	**$75.00**
......	Paperback	0-19-963210-3	**$50.00**
100.	**Diagnostic Molecular Pathology: Volume II: Cell and Tissue Genotyping** Herrington, C.S. & McGee, J.O'D. (Eds)		
......	Spiralbound hardback	0-19-963239-1	**$60.00**
......	Paperback	0-19-963238-3	**$39.00**
99.	**Neuronal Cell Lines** Wood, J.N. (Ed)		
......	Spiralbound hardback	0-19-963346-0	**$68.00**
......	Paperback	0-19-963345-2	**$48.00**
98.	**Neural Transplantation** Dunnett, S.B. & Björklund, A. (Eds)		
......	Spiralbound hardback	0-19-963286-3	**$69.00**
......	Paperback	0-19-963285-5	**$42.00**
97.	**Human Cytogenetics: Volume II: Malignancy and Acquired Abnormalities (2/e)** Rooney, D.E. & Czepulkowski, B.H. (Eds)		
......	Spiralbound hardback	0-19-963290-1	**$75.00**
......	Paperback	0-19-963289-8	**$50.00**
96.	**Human Cytogenetics: Volume I: Constitutional Analysis (2/e)** Rooney, D.E. & Czepulkowski, B.H. (Eds)		
......	Spiralbound hardback	0-19-963288-X	**$75.00**
......	Paperback	0-19-963287-1	**$50.00**
95.	**Lipid Modification of Proteins** Hooper, N.M. & Turner, A.J. (Eds)		
......	Spiralbound hardback	0-19-963274-X	**$75.00**
......	Paperback	0-19-963273-1	**$50.00**
94.	**Biomechanics-Structures and Systems** Biewener, A.A. (Ed)		
......	Spiralbound hardback	0-19-963268-5	**$85.00**
......	Paperback	0-19-963267-7	**$50.00**
93.	**Lipoprotein Analysis** Converse, C.A. & Skinner, E.R. (Eds)		
......	Spiralbound hardback	0-19-963192-1	**$65.00**
......	Paperback	0-19-963231-6	**$42.00**
92.	**Receptor-Ligand Interactions** Hulme, E.C. (Ed)		
......	Spiralbound hardback	0-19-963090-9	**$75.00**
......	Paperback	0-19-963091-7	**$50.00**
91.	**Molecular Genetic Analysis of Populations** Hoelzel, A.R. (Ed)		
......	Spiralbound hardback	0-19-963278-2	**$65.00**
......	Paperback	0-19-963277-4	**$45.00**
90.	**Enzyme Assays** Eisenthal, R. & Danson, M.J. (Eds)		
......	Spiralbound hardback	0-19-963142-5	**$68.00**
......	Paperback	0-19-963143-3	**$48.00**
89.	**Microcomputers in Biochemistry** Bryce, C.F.A. (Ed)		
......	Spiralbound hardback	0-19-963253-7	**$60.00**
......	Paperback	0-19-963252-9	**$40.00**
88.	**The Cytoskeleton** Carraway, K.L. & Carraway, C.A.C. (Eds)		
......	Spiralbound hardback	0-19-963257-X	**$60.00**
......	Paperback	0-19-963256-1	**$40.00**
87.	**Monitoring Neuronal Activity** Stamford, J.A. (Ed)		
......	Spiralbound hardback	0-19-963244-8	**$60.00**
......	Paperback	0-19-963243-X	**$45.00**
86.	**Crystallization of Nucleic Acids and Proteins** Ducruix, A. & Gieg‹130›, R. (Eds)		
......	Spiralbound hardback	0-19-963245-6	**$60.00**
......	Paperback	0-19-963246-4	**$50.00**
85.	**Molecular Plant Pathology: Volume I** Gurr, S.J., McPherson, M.J. & others (Eds)		
......	Spiralbound hardback	0-19-963103-4	**$60.00**
......	Paperback	0-19-963102-6	**$40.00**
84.	**Anaerobic Microbiology** Levett, P.N. (Ed)		
......	Spiralbound hardback	0-19-963204-9	**$75.00**
......	Paperback	0-19-963262-6	**$45.00**

83. **Oligonucleotides and Analogues** Eckstein, F. (Ed)
...... Spiralbound hardback 0-19-963280-4 **$65.00**
...... Paperback 0-19-963279-0 **$45.00**
82. **Electron Microscopy in Biology** Harris, R. (Ed)
...... Spiralbound hardback 0-19-963219-7 **$65.00**
...... Paperback 0-19-963215-4 **$45.00**
81. **Essential Molecular Biology: Volume II** Brown, T.A. (Ed)
...... Spiralbound hardback 0-19-963112-3 **$65.00**
...... Paperback 0-19-963113-1 **$45.00**
80. **Cellular Calcium** McCormack, J.G. & Cobbold, P.H. (Eds)
...... Spiralbound hardback 0-19-963131-X **$75.00**
...... Paperback 0-19-963130-1 **$50.00**
79. **Protein Architecture** Lesk, A.M.
...... Spiralbound hardback 0-19-963054-2 **$65.00**
...... Paperback 0-19-963055-0 **$45.00**
78. **Cellular Neurobiology** Chad, J. & Wheal, H. (Eds)
...... Spiralbound hardback 0-19-963106-9 **$73.00**
...... Paperback 0-19-963107-7 **$43.00**
77. **PCR** McPherson, M.J., Quirke, P. & others (Eds)
...... Spiralbound hardback 0-19-963226-X **$55.00**
...... Paperback 0-19-963196-4 **$40.00**
76. **Mammalian Cell Biotechnology** Butler, M. (Ed)
...... Spiralbound hardback 0-19-963207-3 **$60.00**
...... Paperback 0-19-963209-X **$40.00**
75. **Cytokines** Balkwill, F.R. (Ed)
...... Spiralbound hardback 0-19-963218-9 **$64.00**
...... Paperback 0-19-963214-6 **$44.00**
74. **Molecular Neurobiology** Chad, J. & Wheal, H. (Eds)
...... Spiralbound hardback 0-19-963108-5 **$56.00**
...... Paperback 0-19-963109-3 **$36.00**
73. **Directed Mutagenesis** McPherson, M.J. (Ed)
...... Spiralbound hardback 0-19-963141-7 **$55.00**
...... Paperback 0-19-963140-9 **$35.00**
72. **Essential Molecular Biology: Volume I** Brown, T.A. (Ed)
...... Spiralbound hardback 0-19-963110-7 **$65.00**
...... Paperback 0-19-963111-5 **$45.00**
71. **Peptide Hormone Action** Siddle, K. & Hutton, J.C.
...... Spiralbound hardback 0-19-963070-4 **$70.00**
...... Paperback 0-19-963071-2 **$50.00**
70. **Peptide Hormone Secretion** Hutton, J.C. & Siddle, K. (Eds)
...... Spiralbound hardback 0-19-963068-2 **$70.00**
...... Paperback 0-19-963069-0 **$50.00**
69. **Postimplantation Mammalian Embryos** Copp, A.J. & Cockroft, D.L. (Eds)
...... Spiralbound hardback 0-19-963088-7 **$70.00**
...... Paperback 0-19-963089-5 **$50.00**
68. **Receptor-Effector Coupling** Hulme, E.C. (Ed)
...... Spiralbound hardback 0-19-963094-1 **$70.00**
...... Paperback 0-19-963095-X **$45.00**
67. **Gel Electrophoresis of Proteins (2/e)** Hames, B.D. & Rickwood, D. (Eds)
...... Spiralbound hardback 0-19-963074-7 **$75.00**
...... Paperback 0-19-963075-5 **$50.00**
66. **Clinical Immunology** Gooi, H.C. & Chapel, H. (Eds)
...... Spiralbound hardback 0-19-963086-0 **$69.95**
...... Paperback 0-19-963087-9 **$50.00**
65. **Receptor Biochemistry** Hulme, E.C. (Ed)
...... Spiralbound hardback 0-19-963092-5 **$70.00**
...... Paperback 0-19-963093-3 **$50.00**
64. **Gel Electrophoresis of Nucleic Acids (2/e)** Rickwood, D. & Hames, B.D. (Eds)
...... Spiralbound hardback 0-19-963082-8 **$75.00**
...... Paperback 0-19-963083-6 **$50.00**
63. **Animal Virus Pathogenesis** Oldstone, M.B.A. (Ed)
...... Spiralbound hardback 0-19-963100-X **$68.00**
...... Paperback 0-19-963101-8 **$40.00**
62. **Flow Cytometry** Ormerod, M.G. (Ed)
...... Paperback 0-19-963053-4 **$50.00**
61. **Radioisotopes in Biology** Slater, R.J. (Ed)
...... Spiralbound hardback 0-19-963080-1 **$75.00**
...... Paperback 0-19-963081-X **$45.00**
60. **Biosensors** Cass, A.E.G. (Ed)
...... Spiralbound hardback 0-19-963046-1 **$65.00**
...... Paperback 0-19-963047-X **$43.00**

59. **Ribosomes and Protein Synthesis** Spedding, G. (Ed)
...... Spiralbound hardback 0-19-963104-2 **$75.00**
...... Paperback 0-19-963105-0 **$45.00**
58. **Liposomes** New, R.R.C. (Ed)
...... Spiralbound hardback 0-19-963076-3 **$70.00**
...... Paperback 0-19-963077-1 **$45.00**
57. **Fermentation** McNeil, B. & Harvey, L.M. (Eds)
...... Spiralbound hardback 0-19-963044-5 **$65.00**
...... Paperback 0-19-963045-3 **$39.00**
56. **Protein Purification Applications** Harris, E.L.V. & Angal, S. (Eds)
...... Spiralbound hardback 0-19-963022-4 **$54.00**
...... Paperback 0-19-963023-2 **$36.00**
55. **Nucleic Acids Sequencing** Howe, C.J. & Ward, E.S. (Eds)
...... Spiralbound hardback 0-19-963056-9 **$59.00**
...... Paperback 0-19-963057-7 **$38.00**
54. **Protein Purification Methods** Harris, E.L.V. & Angal, S. (Eds)
...... Spiralbound hardback 0-19-963002-X **$60.00**
...... Paperback 0-19-963003-8 **$40.00**
53. **Solid Phase Peptide Synthesis** Atherton, E. & Sheppard, R.C.
...... Spiralbound hardback 0-19-963066-6 **$58.00**
...... Paperback 0-19-963067-4 **$39.95**
52. **Medical Bacteriology** Hawkey, P.M. & Lewis, D.A. (Eds)
...... Spiralbound hardback 0-19-963008-9 **$69.95**
...... Paperback 0-19-963009-7 **$50.00**
51. **Proteolytic Enzymes** Beynon, R.J. & Bond, J.S. (Eds)
...... Spiralbound hardback 0-19-963058-5 **$60.00**
...... Paperback 0-19-963059-3 **$39.00**
50. **Medical Mycology** Evans, E.G.V. & Richardson, M.D. (Eds)
...... Spiralbound hardback 0-19-963010-0 **$69.95**
...... Paperback 0-19-963011-9 **$50.00**
49. **Computers in Microbiology** Bryant, T.N. & Wimpenny, J.W.T. (Eds)
...... Paperback 0-19-963015-1 **$40.00**
48. **Protein Sequencing** Findlay, J.B.C. & Geisow, M.J. (Eds)
...... Spiralbound hardback 0-19-963012-7 **$56.00**
...... Paperback 0-19-963013-5 **$38.00**
47. **Cell Growth and Division** Baserga, R. (Ed)
...... Spiralbound hardback 0-19-963026-7 **$62.00**
...... Paperback 0-19-963027-5 **$38.00**
46. **Protein Function** Creighton, T.E. (Ed)
...... Spiralbound hardback 0-19-963006-2 **$65.00**
...... Paperback 0-19-963007-0 **$45.00**
45. **Protein Structure** Creighton, T.E. (Ed)
...... Spiralbound hardback 0-19-963000-3 **$65.00**
...... Paperback 0-19-963001-1 **$45.00**
44. **Antibodies: Volume II** Catty, D. (Ed)
...... Spiralbound hardback 0-19-963018-6 **$58.00**
...... Paperback 0-19-963019-4 **$39.00**
43. **HPLC of Macromolecules** Oliver, R.W.A. (Ed)
...... Spiralbound hardback 0-19-963020-8 **$54.00**
...... Paperback 0-19-963021-6 **$45.00**
42. **Light Microscopy in Biology** Lacey, A.J. (Ed)
...... Spiralbound hardback 0-19-963036-4 **$62.00**
...... Paperback 0-19-963037-2 **$45.00**
41. **Plant Molecular Biology** Shaw, C.H. (Ed)
...... Paperback 1-85221-056-7 **$38.00**
40. **Microcomputers in Physiology** Fraser, P.J. (Ed)
...... Spiralbound hardback 1-85221-129-6 **$54.00**
...... Paperback 1-85221-130-X **$36.00**
39. **Genome Analysis** Davies, K.E. (Ed)
...... Spiralbound hardback 1-85221-109-1 **$54.00**
...... Paperback 1-85221-110-5 **$36.00**
38. **Antibodies: Volume I** Catty, D. (Ed)
...... Paperback 0-947946-85-3 **$38.00**
37. **Yeast** Campbell, I. & Duffus, J.H. (Eds)
...... Paperback 0-947946-79-9 **$36.00**
36. **Mammalian Development** Monk, M. (Ed)
...... Hardback 1-85221-030-3 **$60.00**
...... Paperback 1-85221-029-X **$45.00**
35. **Lymphocytes** Klaus, G.G.B. (Ed)
...... Hardback 1-85221-018-4 **$54.00**
34. **Lymphokines and Interferons** Clemens, M.J., Morris, A.G. & others (Eds)
...... Paperback 1-85221-035-4 **$44.00**
33. **Mitochondria** Darley-Usmar, V.M., Rickwood, D. & others (Eds)
...... Hardback 1-85221-034-6 **$65.00**
...... Paperback 1-85221-033-8 **$45.00**

32.	Prostaglandins and Related Substances Benedetto, C., McDonald-Gibson, R.G. & others (Eds)		
......	Hardback	1-85221-032-X	$58.00
......	Paperback	1-85221-031-1	$38.00
31.	DNA Cloning: Volume III Glover, D.M. (Ed)		
......	Hardback	1-85221-049-4	$56.00
......	Paperback	1-85221-048-6	$36.00
30.	Steroid Hormones Green, B. & Leake, R.E. (Eds)		
......	Paperback	0-947946-53-5	$40.00
29.	Neurochemistry Turner, A.J. & Bachelard, H.S. (Eds)		
......	Hardback	1-85221-028-1	$56.00
......	Paperback	1-85221-027-3	$36.00
28.	Biological Membranes Findlay, J.B.C. & Evans, W.H. (Eds)		
......	Hardback	0-947946-84-5	$54.00
......	Paperback	0-947946-83-7	$36.00
27.	Nucleic Acid and Protein Sequence Analysis Bishop, M.J. & Rawlings, C.J. (Eds)		
......	Hardback	1-85221-007-9	$66.00
......	Paperback	1-85221-006-0	$44.00
26.	Electron Microscopy in Molecular Biology Sommerville, J. & Scheer, U. (Eds)		
......	Hardback	0-947946-64-0	$54.00
......	Paperback	0-947946-54-3	$40.00
25.	Teratocarcinomas and Embryonic Stem Cells Robertson, E.J. (Ed)		
......	Hardback	1-85221-005-2	$62.00
......	Paperback	1-85221-004-4	$0.00
24.	Spectrophotometry and Spectrofluorimetry Harris, D.A. & Bashford, C.L. (Eds)		
......	Hardback	0-947946-69-1	$56.00
......	Paperback	0-947946-46-2	$39.95
23.	Plasmids Hardy, K.G. (Ed)		
......	Paperback	0-947946-81-0	$36.00
22.	Biochemical Toxicology Snell, K. & Mullock, B. (Eds)		
......	Paperback	0-947946-52-7	$40.00
19.	Drosophila Roberts, D.B. (Ed)		
......	Hardback	0-947946-66-7	$67.50
......	Paperback	0-947946-45-4	$46.00
17.	Photosynthesis: Energy Transduction Hipkins, M.F. & Baker, N.R. (Eds)		
......	Hardback	0-947946-63-2	$54.00
......	Paperback	0-947946-51-9	$36.00
16.	Human Genetic Diseases Davies, K.E. (Ed)		
......	Hardback	0-947946-76-4	$60.00
......	Paperback	0-947946-75-6	$34.00
14.	Nucleic Acid Hybridisation Hames, B.D. & Higgins, S.J. (Eds)		
......	Hardback	0-947946-61-6	$60.00
......	Paperback	0-947946-23-3	$36.00
13.	Immobilised Cells and Enzymes Woodward, J. (Ed)		
......	Hardback	0-947946-60-8	$0.00
12.	Plant Cell Culture Dixon, R.A. (Ed)		
......	Paperback	0-947946-22-5	$36.00
11a.	DNA Cloning: Volume I Glover, D.M. (Ed)		
......	Paperback	0-947946-18-7	$36.00
11b.	DNA Cloning: Volume II Glover, D.M. (Ed)		
......	Paperback	0-947946-19-5	$36.00
10.	Virology Mahy, B.W.J. (Ed)		
......	Paperback	0-904147-78-9	$40.00

9.	Affinity Chromatography Dean, P.D.G., Johnson, W.S. & others (Eds)		
......	Paperback	0-904147-71-1	$36.00
7.	Microcomputers in Biology Ireland, C.R. & Long, S.P. (Eds)		
......	Paperback	0-904147-57-6	$36.00
6.	Oligonucleotide Synthesis Gait, M.J. (Ed)		
......	Paperback	0-904147-74-6	$38.00
5.	Transcription and Translation Hames, B.D. & Higgins, S.J. (Eds)		
......	Paperback	0-904147-52-5	$38.00
3.	Iodinated Density Gradient Media Rickwood, D. (Ed)		
......	Paperback	0-904147-51-7	$36.00

Sets

	Essential Molecular Biology: Volumes I and II as a set Brown, T.A. (Ed)		
......	Spiralbound hardback	0-19-963114-X	$118.00
......	Paperback	0-19-963115-8	$78.00
	Antibodies: Volumes I and II as a set Catty, D. (Ed)		
......	Paperback	0-19-963063-1	$70.00
	Cellular and Molecular Neurobiology Chad, J. & Wheal, H. (Eds)		
......	Spiralbound hardback	0-19-963255-3	$133.00
......	Paperback	0-19-963254-5	$79.00
	Protein Structure and Protein Function: Two-volume set Creighton, T.E. (Ed)		
......	Spiralbound hardback	0-19-963064-X	$114.00
......	Paperback	0-19-963065-8	$80.00
	DNA Cloning: Volumes I, II, III as a set Glover, D.M. (Ed)		
......	Paperback	1-85221-069-9	$92.00
	Molecular Plant Pathology: Volumes I and II as a set Gurr, S.J., McPherson, M.J. & others (Eds)		
......	Spiralbound hardback	0-19-963354-1	$0.00
......	Paperback	0-19-963353-3	$0.00
	Protein Purification Methods, and Protein Purification Applications, two-volume set Harris, E.L.V. & Angal, S. (Eds)		
......	Spiralbound hardback	0-19-963048-8	$98.00
......	Paperback	0-19-963049-6	$68.00
	Diagnostic Molecular Pathology: Volumes I and II as a set Herrington, C.S. & McGee, J. O'D. (Eds)		
......	Spiralbound hardback	0-19-963241-3	$0.00
......	Paperback	0-19-963240-5	$0.00
	Receptor Biochemistry; Receptor-Effector Coupling; Receptor-Ligand Interactions Hulme, E.C. (Ed)		
......	Spiralbound hardback	0-19-963096-5	$193.00
......	Paperback	0-19-963097-6	$125.00
	Signal Transduction Milligan, G. (Ed)		
......	Spiralbound hardback	0-19-963296-0	$60.00
......	Paperback	0-19-963295-2	$38.00
	Human Cytogenetics: Volumes I and II as a set (2/e) Rooney, D.E. & Czepulkowski, B.H. (Eds)		
......	Hardback	0-19-963314-2	$130.00
......	Paperback	0-19-963313-4	$90.00
	Peptide Hormone Secretion/Peptide Hormone Action Siddle, K. & Hutton, J.C. (Eds)		
......	Spiralbound hardback	0-19-963072-0	$135.00
......	Paperback	0-19-963073-9	$90.00

ORDER FORM for USA and Canada

Qty	ISBN	Author	Title	Amount
			S&H	
	CA and NC residents add appropriate sales tax			
			TOTAL	

Please add shipping and handling: $2.50 for first book, ($1.00 each book thereafter)

Name ....

Address ....

....

.... Zip ....

[ ] Please charge $ .... to my credit card
Mastercard/VISA/American Express (circle appropriate card)

Acct. .... Expiry date ....

Signature ....

Credit card account address if different from above:

....

.... Zip ....

[ ] I enclose a cheque for $............

Mail orders to: Order Dept. Oxford University Press, 2001 Evans Road, Cary, NC 27513